THE SUN

# THE SUN

## PART I
## OF
## SOLAR-TERRESTRIAL PHYSICS/1970

COMPRISING THE PROCEEDINGS OF THE
INTERNATIONAL SYMPOSIUM ON SOLAR-TERRESTRIAL PHYSICS
HELD IN LENINGRAD, U.S.S.R.
12–19 MAY 1970

*Sponsored by COSPAR, IAU, IUGG-IAGA, and URSI*

C. DE JAGER
*Editor*

E. R. DYER
*General Editor of the Proceedings*

D. REIDEL PUBLISHING COMPANY

DORDRECHT-HOLLAND

## TABLE OF CONTENTS

# SOLAR ENERGY SOURCES

C. DE JAGER
*The Astronomical Institute, Utrecht, The Netherlands*

**Abstract.** The energy content of the sun is considered, and mechanisms for the dissipation and emission of the various kinds of energy are examined. The most important energy flux is the radiant flux originating from nuclear reactions in the solar interior. In the outermost $10^5$ km of the solar body this flux is transported convectively. On top of the convective layer, a region of about 600 km thick constituting the photosphere and low chromosphere is in near-radiative equilibrium. The structure of the chromosphere and corona is mainly due to a mechanical energy flux emanating from the outer convective region. The various transient solar phenomena, which for a part involve high energy processes in magnetic regions, demand thermodynamic upgrading of energy, which, in general terms, can only occur in a motion field with locally a non-zero curl, i.e. solar differential rotation or convection. It is shown that the secular decrease of solar rotational energy can be due to the emission of transformed rotational energy by flares. It is suggested that this happens through the intermediary of differential rotation, which in turns originates by coupling of the solar rotation with convection (see Figure 1).

## 1. The Basic Problem of Solar-Terrestrial Physics

The basic problem of solar-terrestrial physics is to understand the interaction between two celestial bodies, through the emission of particles, magnetic fields and electromagnetic radiation from one to the other.

(a) The first body is a cloud of gas, mass $2 \times 10^{33}$ gram, composition by mass 60% H, 40% He, and less than 1% heavier particles, with a given age ($5 \times 10^9$ years), angular momentum, and magnetic moment.

(b) The other body, at a distance of $1.5 \times 10^8$ km, is diamagnetic but has a known nearly poloidal magnetic field, has an atmosphere, with a given mass and composition.

The above boundary values of the problem would suffice to establish the complete picture of solar-terrestrial interactions. If we would fully understand it all, the data would allow one to predict why a certain flare would originate at a certain moment with its particular emissions, and one should be able to understand what is happening in the solar atmosphere, interplanetary space, in the magnetosphere and ionosphere. Yet the actual situation is different, and our insufficient knowledge of the application of basic physics to astrophysical problems, let it be our lack of imagination, forces us to build up the picture, as we actually do, from the observational side. The observations are apparently the stepping stones that allow us to cross the river of our ignorance, and to build up an as yet incoherent picture of solar-terrestrial relations.

## 2. Solar Energies

The sun has the following energy contents:

*Gravitational energy* being the energy acquired during the contraction of solar

*Dyer (ed.), Solar Terrestrial Physics / 1970: Part I*, 1–8. *All Rights Reserved.*

matter from infinity: $E_G = bGM^2/R$, where $G$ is the gravitation constant, $M$ the solar mass $= 2 \times 10^{33}$ g and $R$ its radius $= 7 \times 10^{10}$ cm; $b$ is a factor of the order unity, dependent on the internal mass distribution. With a reasonable assumption about $b$, based on model computations, one finds $E_G = 8 \times 10^{48}$ erg.

*Thermal or translation energy* $E_T = \int NkT \,\mathrm{d}V$, where $N$ is the particle number density, and $\mathrm{d}V$ a volume element. $E_T = 3 \times 10^{48}$ erg.

*The excitation and ionization energies*

$$E_{EI} = \int N \sum \nu_i \left\{ \sum_{j=1}^{i} x_{ij} \left( \chi_{ij} + \sum_{e=1}^{\infty} x_{ije}\chi_{ije} \right) \right\} \mathrm{d}V ,$$

where $\nu_i$ is the fractional number of particles of species $i$, $x_{ij}$ the degree of ionization ($j$) of the particles $i$, and $\chi_{ij}$ the ionization energy; $\chi_{ije}$ is the excitation energy of particles $i$ in the state of ionization $j$ and of excitation $e$. Generally one defines $E_T + E_{EI} = E_i$, the internal energy. Further $E_i + E_G$ is the total supranuclear energy.

*Hydrogenic nuclear energy* $E_H = \Delta c^2 \int_0^{M_1} X \,\mathrm{d}m$ where $\Delta$ is the mass defect of alpha particles with respect to four times the number of protons, per gram of protons ($\Delta = 0.0072$ g/g); $X$ the solar mass fraction of hydrogen, $\mathrm{d}m$ a mass element, and $M_1$ the mass of that part of the solar interior where nuclear reactions can occur; roughly $M_1 \approx 0.1\, M$. With $X = 0.5$, $E_H = 6 \times 10^{51}$ erg.

*Rotational energy* $E_R = \int_0^M r^2\omega^2 \,\mathrm{d}m = 2 \times 10^{42}$ erg, where $r$ is the distance between the mass element $\mathrm{d}m$ and the axis of rotation, $\omega$ is the angular velocity.

*The energy of differential rotation* $E_{DR} = \int_0^M r^2(\omega - \varpi)^2 \,\mathrm{d}m$, needs some evaluation, since it depends on the way angular rotation is distributed through the solar body. On the basis of various recent models of the solar inner rotation $E_{DR} \approx 2 \times 10^{40}$ erg may appear to be a good average value.

*The magnetic energy* $E_M = (1/4\pi) \int B^2 \,\mathrm{d}V$ can not yet be estimated with any reasonable certainty.

*The subnuclear energy* $Mc^2 = 2 \times 10^{54}$ erg is not important in the present context.

## 3. Energy Fluxes

If the above described energies would remain constant, the sun would be invisible to any observer with any observational technique.

The way the sun interacts with the surrounding celestial bodies like the earth, is by *dissipation* of any of the above given energies. This dissipation yields *energy fluxes* $F = \mathrm{d}E/\mathrm{d}t$, that will next be estimated. The most important flux is the *radiant flux* $F_R$, due to thermonuclear H→He reactions in the solar interior. Observationally:

$$F_R = 3.8 \times 10^{33} (\pm 2\%) \text{ erg sec}^{-1} ,$$

which corresponds to a flux per cm$^2$: $f_R = 6 \times 10^{10}$ erg cm$^{-2}$ sec$^{-1}$. If the sun would contract, its gravitational energy $E_G$ would decrease and $E_i$ and $E_T$ would change correspondingly. By a simple thermodynamic reasoning it can be shown that

$$\mathrm{d}E_T = \mathrm{d}E_i + \mathrm{d}E_G = \frac{3\gamma - 4}{3(\gamma - 1)}\,\mathrm{d}E_G\,,$$

where $\gamma$ is the ratio of specific heats of matter. With $\gamma = \frac{5}{3}$ (mono-atomic gases), one obtains $\mathrm{d}E_T = \frac{1}{2}\mathrm{d}E_G$, so that half of the lost gravitational energy would be emitted as radiation. However, in its actual phase of evolution the sun would slightly *expand* rather than contract, but only so slowly that the corresponding fluxes are wholly unimportant for the solar energy balance. So we may neglect the changes of the gravitation, thermal, excitation and ionization energy contents of the sun as sources of emitted energy.

Rotational energy can decrease by viscous losses due to non-uniform rotation of the solar body, or through its transformation into magnetic energy and subsequent annihilation or emission of magnetic fields. These losses of rotational and magnetic energy (by neutralization or by dragging out of magnetic fields by expelled gases) will be discussed in Section 5.

## 4. The Radiation Flux; Structure of the Outer Solar Convection Zone and Photosphere

The solar radiation flux $F_R$ may propagate outward in two modes: convectively and radiatively. The mode that actually occurs in a certain part of the solar body depends on which of the two absolute values of the temperature gradient is the smaller:

*the radiative gradient*: $$\left(\frac{\mathrm{d}T}{\mathrm{d}r}\right)_{\mathrm{rad}} = \frac{3\kappa\varrho}{16\sigma T}\frac{F_R}{4\pi r^2}\,,$$

or

*the convective gradient*: $$\left(\frac{\mathrm{d}T}{\mathrm{d}r}\right)_{\mathrm{conv}} = \left(1 - \frac{1}{\gamma}\right)\frac{T}{P}\frac{\mathrm{d}P}{\mathrm{d}r}\,,$$

where $\kappa$ is the absorption coefficient of solar matter, and $\varrho$ the mass density; $\sigma$ is Stefan-Boltzmann's constant; $\gamma$ is the ratio of specific heats of matter. The importance of convection is that it produces a system of up and downward motions, hence a motion field with locally a non-zero curl. Therefore a convective region is a source of waves, that can give rise to non-thermal effects like the heating of the solar corona, and that may also lead to other suprathermal effects of wave-gas interaction.

There seems to be a small convective region in the very solar centre, but this may have no direct influence on the sun's outer parts.

The travel time of a sound wave from the solar centre to its surface $t_s = \int_0^R \mathrm{d}r/v_s$ is of the order of a few hours, but it would be completely damped on its way out.

A radiative disturbance would take a time of

$$t_r = \int_0^R \kappa\tau\beta\,\mathrm{d}r$$

where $\kappa^{-1}$ is the mean free path of a photon, $\tau$ the average lifetime of an excited state and $\beta$ a multiplication term describing the effect of the brownian motions of photons propagating through the solar interior. With reasonable assumptions ($\kappa = 10^{-3}$ cm$^{-1}$, $\tau = 10^{-8}$ sec; $\beta > 10$) one obtains $t_r > 7 \times 10^6$ sec, i.e. more than 100 days. Solar core flashes, should they ever exist, would be fully smeared out in time. Henceforth in this paper we shall disregard the central convective region and its possible influence on the outer layers.

The outer convective region is of far greater importance. It extends from a depth of about $10^5$ km up to very close to the surface and is manifest in the well-known solar *granulation* pattern. Actually photospheric observations show the convective motions to extend up to a level, about 200 km below the limb surface (monochromatic optical depth at 5000 Å: $\tau_5 = \int_R^\infty \kappa_5 \, dr \approx 0.1$). In the uppermost 100 km of the convection zone the observed average up- and downward velocities are about 2 km/sec in fair agreement with theoretical predictions; they decrease rapidly in the denser deeper regions. Notwithstanding these large convective velocities the uppermost 100 or 200 km of the convection zone are in near radiative equilibrium i.e. energy is transported radiatively there, which is due to the large mean free paths of the photons in this already very transparent part of the sun.

The radiative domain extends upward to $\tau_5 \approx 10^{-4}$, i.e. a level about 200 km above the limb level. Hence the zone that is nearly completely in radiative equilibrium occurs between monochromatic optical depths of approximately $\tau_5 = 10^{-4}$ and 2, and has a total thickness of about 600 km.

In this whole range of depths the temperature distribution is defined by the laws of radiative equilibrium, but deviations occur at the two extremities. Near $\tau_5 = 10^{-5}$ a temperature inversion occurs and higher upward the temperature increases steeply. This is due to dissipation of mechanical energy; waves emerging from the convection zone and visible as large-scale vibrations of the upper photosphere and chromosphere, propagate upward into the more and more tenuous outer solar layers. They carry a mechanical energy flux $f_m = \theta \varrho v^2 v_s$, where $v$ is the velocity amplitude of the waves, $v_s$ the velocity of sound, and $\theta$ a quantity of the order unity. Since the mechanical flux in the lowest chromospheric levels is constant with height the velocity amplitude initially increases with decreasing $\varrho$. By the effect of second order terms in the wave equation the waves, when propagating upward, lose their sinusoidal character and change into shockwaves which dissipate energy. Otherwise stated: higher harmonics with larger frequencies are generated: these are more efficient in dissipating energy.

The temperature in the region above $\tau_5 = 10^{-4}$ is nearly completely defined by the dissipation of the mechanical energy flux by viscous losses in shock waves whereas the main radiative flux passes without being absorbed. Hence we may call this upper region the mechanical domain in contrast to the radiative and convective regions below it. The mechanical flux as deduced from a comparison of the observed and theoretically predicted temperature distributions near $\log \tau_5 = -4$ is about $2 \times 10^6$ erg cm$^{-2}$ sec$^{-1}$ hence $3 \times 10^{-5}$ times the radiative flux.

We compare this flux with the total wave energy of a column with 1 cm$^2$ cross section of the solar convective zone.

$$E_c = \tfrac{1}{4}\int_{r_1}^{r_2} \varrho v^2 \, \mathrm{d}r = \frac{\ln 10}{4g}\int_{r_1}^{r_2} v^2 P_g \, \mathrm{d}(\log P_g) \approx 10^{13} \text{ erg cm}^{-2} .$$

Here $r_1$ and $r_2$ are the lower and upper limits of the convection zone, and $v$ is the velocity amplitude; $g$ is the acceleration of gravity. A comparison of $E_c$ with $f_m$ shows that the dissipation efficiency of this energy is approximately $2 \times 10^{-7}$ sec$^{-1}$. Yet this relatively small amount of energy is sufficient for yielding the whole chromosphere, corona and solar wind.

The various energies discussed in this paper are summarized in Table I.

TABLE I

Solar energies

| Energy | |
|---|---|
| Gravitational | $E_G = 8 \times 10^{48}$ erg |
| Thermal | $E_T = 3 \times 10^{48}$ erg |
| Rotational | $E_R = 2 \times 10^{42}$ erg |
| Differential rotation | $E_{DR} = 2 \times 10^{40}$ erg |
| Wave energy of convection | $E_C = 6 \times 10^{35}$ erg |
| Energy of supergranulation | $E_{SG} = 10^{34}$ erg |
| Nuclear (hydrogen) | $E_H = 6 \times 10^{51}$ erg |
| Subnuclear | $E_S = 2 \times 10^{54}$ erg |

## 5. The Origin of Solar Variability

First we have to determine the source of the energy of the solar transient phenomena. Since the most important of these, like solar flares, involve particle energies ranging from 1 to $10^9$ eV, the mechanism seems to demand at least partial thermodynamical upgrading of energy. It may therefore seem that the purely radiative flux, although being the largest of all should be excluded, since the only way to upgrade energy seems to be through rotational or magnetic phenomena, hence by the introduction of waves or by magnetohydrodynamic processes. However, it will be shown that the convection zone, where the main solar energy flux is transported convectively may contribute to this upgrading process.

TABLE II

Emitted flare energies

| Component | Characteristic particle energy $E$ | Average particle density (cm$^{-3}$) | Total number of particles involved $\int N \mathrm{d}V$ | Emitted or carried total energy (erg) |
|---|---|---|---|---|
| Optical flare plasma | 2 eV | $3 \times 10^{13}$ | $3 \times 10^{36}$ | $10^{31}$ |
| Interplanetary plasma front | 0.5 keV | – | $10^{39}$ | $10^{30}$ |
| High energy flare plasma | 20 keV | $10^{10}$ | $10^{38}$ | $10^{30}$ |
| (Sub-)relativistic particle component | 0.2 GeV | – | $2 \times 10^{33}$ | $10^{30}$ |

We wish to compare the energy contents listed in Table I with the energy dissipated in solar flares. In Table II we give values derived by us for a medium-type flare, of importance 2. The table makes clear that for an average flare an average energy of $10^{31}$ erg is emitted or fed into plasmas of greatly varying average particle energy. For weak or strong flares (importances 1 and 4 respectively) the numbers may be smaller or larger, approximately by factors 10 to $10^2$.

Assuming that the average energy emitted during one average solar cycle is a few thousand times the total energy emitted by a type 2 flare, hence a few times $10^{34}$ erg, one obtains an average rate of dissipation of flare energy of a few times $10^{34}/3 \times 10^8$ $\approx 10^{26}$ erg $\sec^{-1}$. This and other energy fluxes are given in Table III.

TABLE III

Solar energy fluxes

| | | |
|---|---|---|
| Radiative flux | $F_{\mathrm{rad}}$ | $= 4 \times 10^{33}$ erg sec$^{-1}$ |
| Mechanical flux at top of photosphere | $F_{\mathrm{mech}}$ | $= 10^{29}$ erg sec$^{-1}$ |
| Coronal X-ray flux | $F_X$ | $= 10^{27}$ erg sec$^{-1}$ |
| Flux of thermal energy in the solar wind | $F_{\mathrm{wind}}$ | $= 10^{25}$ erg sec$^{-1}$ |
| Dissipation of rotational energy by the solar wind | $F_{\mathrm{rot,\,wind}}$ | $= 10^{24}$ erg sec$^{-1}$ |
| Secular decrease of solar rotational energy | $F_R$ | $= 8 \times 10^{25}$ erg sec$^{-1}$ |
| Average energy dissipation in flares | $F_{\mathrm{flares}}$ | $= 10^{26}$ erg sec$^{-1}$ |

It may seem surprising that the value of $\bar{E} \int N \mathrm{d}V$ derived from this table for the optical flare plasma is $10^{25}$ erg, hence about $10^6$ times smaller than the total emitted energy (last column) but this simply shows that during the life time of the optical flare there is a continuous re-excitation of the particles in the cool flare plasma (actually $10^6$ times per particle during the life time of the optical flare).

The most essential problem of solar flares is how individual particles can acquire the high particle energies, given in the second column of Table II.

We also have to observe the time scale: a flare may recur in a series of homologous events with time intervals of the order of a day ($10^5$ sec). On the other hand it seems after H. Dodson-Prince and R. Hedeman (in this volume, p. 151), that it may take an active region a couple of years before sufficient energy is accumulated to produce real high particle energy flares. Another important aspect of flares is that their origin is always related to electric currents, of at least $10^{11}$ ampères. Data about particle energies and total energy, time rate and electric currents are necessary to define the kind of flare producing mechanisms.

We examine the hypothesis that the origin of solar active regions and flares is due, in one way or the other to solar rotation and its differential effects. In the non-magnetic case the dissipation mechanism would just be viscosity, but in the magnetic case differential rotation of a magnetized plasma would produce a current system

$$\mathbf{j} = \frac{c}{4\pi} \operatorname{curl} \mathbf{B},$$

which would upgrade and dissipate energy by an instability leading to current interruption and large electric fields. It is mostly assumed, though never proved exactly that in many turbulent plasma regions, or more generally in regions for which

$$\int |\mathrm{curl}\,\mathbf{v}|\,\mathrm{d}V \neq 0,$$

eventually a certain degree of equipartition is reached between mechanical and magnetic energies. We shall assume this here, and consider further only magnetic field dissipation since the considered manifestations of solar activity are restricted to magnetic regions. We take further that the sun dissipates flare energy at an average rate of $10^{26}$ erg sec$^{-1}$. Hence, assuming flares to be due to magnetohydrodynamic dissipation of solar differential rotation energy one would find that the presently available *differential* rotational energy of the sun ($2 \times 10^{40}$ erg) would be exhausted in approximately $10^6$ to $10^7$ cycles, $\approx 10^7$ to $10^8$ yr. This seems impossibly short as compared with the time during which the sun must have had its present properties.

It is, however, incorrect to compare the rate of dissipation of flare energy with the total energy content of *differential* rotation since the differential rotation is the result of a coupling between the average solar rotation (or: the quasi solid body-rotation of the core) with the outer convective region (de Jager, 1959; Durney, 1970; Roxburgh, 1970). Hence, loss of the energy of differential solar rotation by the flare effect can be replenished from the total energy of solar rotation, so that we have to consider the total energy content of the average solar rotation, and its dissipation.

According to Conti and van den Heuvel (1971) the *e*-folding time of the average solar rotation is approximately $1.5 \times 10^9$ year. The *e*-folding time of the total rotational *energy* is half this value, $\approx 8 \times 10^8$ year $= 2.4 \times 10^{16}$ sec. Hence, the actual average dissipation of solar rotational energy is $2 \times 10^{42}/2.4 \times 10^{16} = 8 \times 10^{25}$ erg sec$^{-1}$. This value is so close to the estimated average energy loss by flares ($\approx 10^{26}$ erg sec$^{-1}$) that it seems tempting to assume that the secular decrease of the rotational energy of the sun may be completely due to the transformation of the energy of differential solar rotation into magnetic energy and its subsequent dissipation in flares. Otherwise stated: we shall assume the dissipation of the energy of solar rotation to be the main source of flare energy.

It should yet be examined by what mechanism the energy of solar rotation is fed into flares. Apparently this mechanism should be related to the systematic motions that may occur in active regions. Quite tentatively the question may be asked whether flares originate in active regions under the influence of the outward streaming motion in the supergranulation. The total kinetic energy content of the streaming motion in one supergranule is $10^{29}$ to $10^{30}$ erg. The magnetic energy content in the boundary of a supergranule is $2\pi r \Delta r z B^2/4\pi$, where $r$ and $\Delta r$ are the radius (20000 km) and boundary thickness (5000 km) of a supergranule and $z$ its depth (1000 km?), $B$ is the magnetic field, $\approx 100$ G. Hence, $E_m = 5 \times 10^{29}$ erg. The total energy content of a supergranule is therefore $10^{30}$ erg; integrated over the surface $10^{34}$ erg.

The life time of a supergranule is of the order of 20 h. This fact, combined with the observation that the energy content of an average supergranule is only a factor

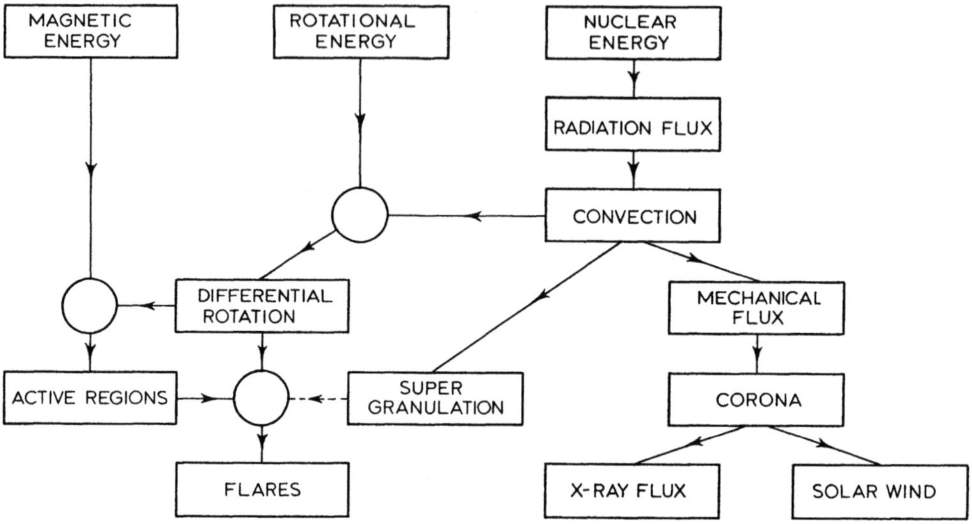


Fig. 1. Schematic representation of the solar energy sources, their interaction and the origin of high energy phenomena like flares and the solar corona. Circles denote coupling or interaction between different phenomena. The primordial part, solar convection plays for solar high energy phenomena is obvious: it leads on one hand, by dissipation of wave energy to the origin of the solar corona, and on the other hand, by coupling with the rotational energy to differential rotation; this in turn, by interaction with magnetic regions, leads to flares. The contribution of supergranular phenomena to flares is still an open problem.

10 smaller than the energy radiated in a flare, leads to the suggestion that the coupling of supergranular features with solar active regions may locally lead to the accumulation of 10 times larger energies, which, by one way or the other are dissipated in flares, which can recur in about one day. Remains then to be solved how the streaming motion in supergranules or in similar structures in active regions can give rise to the observed current system that eventually leads to a flare.

## Acknowledgement

I wish to thank Professor E. van den Heuvel for a very stimulating discussion and interesting suggestions on the problem of solar and stellar rotation.

## References

Conti, P. S. and van den Heuvel, E.: 1971, in preparation.
De Jager, C.: 1959, *Handbuch der Physik* **52**, 343.
Durney, B. R.: 1970, in A. Slettebak (ed.), *Stellar Rotation*, D. Reidel, Dordrecht, p. 30.
Hundhausen, A. J.: 1968, *Space Sci. Rev.* **8**, 690.
Roxburgh, I. W.: 1970, in A. Slettebak (ed.), *Stellar Rotation*, D. Reidel, Dordrecht, p. 318.

# STRUCTURE AND DYNAMICS OF THE SOLAR CORONA

M. KUPERUS

*The Astronomical Institute, University of Utrecht, The Netherlands*

**Abstract.** The structure of the outer solar atmosphere will be discussed in the context of the processes of energy transfer that play a part in the solar corona. Special emphasis is laid on the competitive role played by the four main energy processes, dissipation, radiation, heat conduction, and hydrodynamic expansion.

A steadily heated quiet outer solar atmosphere may be divided into several rather distinct layers according to the dominance of one or more of the processes mentioned before.

The influence of solar activity and magnetic fields on the structure and energy balance of the corona is discussed, particularly the consequences for the heat flow and solar wind in coronal streamers.

## 1. Introduction

The structure and the physical state of the outer layers of the sun, i.e., the chromosphere and corona, can be revealed only from a combination of observations over a spectral range that almost covers the electromagnetic spectrum ranging from X-rays to radiowaves. Additional information may be gathered from the continuous low-energy corpuscular radiation of the sun, the so-called solar wind. None of these components is really constant in time and most of the radiation seems to come from small regions on the sun that are strongly associated with solar active regions.

One does not expect a medium that emits so irregularly in such a broad spectral range to be described by simple physics; neither does one have much hope of being able to represent the coronal plasma by a few simple state parameters.

The problem of the study of the corona is threefold. First, there are the difficulties in obtaining adequate observations. Until recently a solar eclipse provided the only possibility (and it is still the best possibility) of observing the corona in visible wavelengths. For the X-ray and corpuscular emission, spaceborne observations are to be made. The spatial resolution changes tremendously through the spectral range. Second, there is the problem of interpreting the observations and of determining the physical parameters and structure of the atmosphere. The third problem is the theory of the corona and the construction of a model atmosphere which takes into account the various energy processes that play a role.

In this review special attention will be paid to the mechanism of energy transport in the outer layers and the consequences for the structure of the solar corona. For a detailed analysis of the observational techniques and a critical evaluation of the coronal data, we refer to the existing reviews in this field (Billings, 1966; Newkirk, 1967) and the earlier contributions of Van de Hulst (1953) and Shklovskiy (1965).

We will briefly mention some of the essentials, since an extensive discussion of the observational aspects would go far beyond the scope of this contribution.

*Dyer (ed.), Solar Terrestrial Physics / 1970: Part I, 9–20.* 

The brightest contribution to the coronal light is the white light corona caused by photospheric light scattered by free electrons. White light observations yield the overall density structure of the corona up to a distance of about 10 solar radii. From the occultations of radio sources the electron density has been found to a distance of 80 $R_{\odot}$. At still greater distances, up to 1 AU, the density has been deduced from spacecraft measurements. Coronal photographs show that the corona is far from isotropic. It is well known that the so-called shape of the corona reflects the same 11-year cycle as solar activity. There are a great number of different inhomogeneous structures all characterized by a greater brightness and hence by an electron density enhancement. Streamers, helmets, arches, active region enhancements, coronal condensations and polar plumes are all examples of inhomogeneous structure which eventually determine the overall 'shape' of the corona. The density enhancement above active regions is of the order of a factor 10. In streamers it may be somewhat larger at greater distances.

The temperature structure is more difficult to determine. In general one can determine the temperature (1) from the observed density distribution, under the assumption that the atmosphere is in hydrostatic equilibrium; (2) from the intensities of the forbidden lines and the theory of ionization equilibrium, and (3) from the intensity of the continuous radio radiation. All these observations now yield more or less consistent values, around $T=1.5\times10^6$ K to $2\times10^6$ K. In the active region enhancements the temperature is somewhat higher, sometimes from $T=3\times10^6\,\mathrm{K}-5\times10^6\,\mathrm{K}$ (Billings, 1966). The coronal temperature gradually declines from its maximum to a value of the order of $10^5$ K at a distance of 1 AU (Neugebauer and Snyder, 1966). The spatial variations in temperature at great distances seem to be much larger than those close to the sun.

The temperature distribution is of great importance as a check of the theory of the solar wind. Actually the way in which the energy is transported outward is reflected much better in the temperature distribution than it is in the density distribution.

The third important physical parameter, besides density and temperature, is the magnetic field distribution. Only very few measurements of coronal fields exist. All observations refer to the coronal active regions and, except for prominences, they are not direct measurements but semitheoretical determinations such as measurements obtained from the oscillations of filaments and the polarization of radiobursts. According to Newkirk (1967) the magnetic field strengths are of the order of 1 to 100 gauss at distances of less than one solar radius from the surface, diminishing to values of the order of $10^{-4}$–$10^{-5}$ gauss at the earth's orbit.

On the other hand in the inner parts of the corona it seems possible to get a satisfactory agreement between the coronal field configuration determined from the observed photospheric magnetic fields and the morphological structure observed during a solar eclipse. For the 1966 eclipse Altschuler and Newkirk (1969) showed that a potential field distribution below a level of 2.5 $R_{\odot}$ is consistent with observations of the structure of the corona. This would mean that the density structure would follow closely the magnetic fields.

## 2. The Energy Balance in the Corona

The amount of heat required to maintain the corona can be found from an estimate of the losses. There are two processes that transport energy from the corona into interstellar space, viz. electromagnetic radiation and the solar wind. In the inner corona thermal conduction constitutes the dominant means of energy loss, but convection by the solar wind gradually takes over in the outer corona. The total radiative losses above a level in the corona where the electron density $n_e \approx 10^9$ cm$^{-3}$ is about $5 \times 10^4$ erg/cm$^2$ sec. The same amount of energy is convected by the solar wind (Kuperus, 1969).

One could easily conclude that the two main kinds of energy loss are comparable. However, the radiative losses increase with the square of the density. Consequently, the radiative losses are strongly dominant in the lowest parts of the corona and in those regions where the density is locally enhanced.

At coronal temperatures hydrogen and helium are completely ionized and thermal conduction is established by the electrons. Therefore the coefficient of thermal conduction is extremely large. This means that, from the level of maximum coronal temperature, energy is conducted both inward into the chromosphere through the transition region and outward into the interplanetary medium. The outward flux of energy is somewhat smaller than the solar wind energy flux. It seems that the greater part of this conductive flux is used to drive the solar wind beyond the levels where the dissipation of mechanical energy is the main heat source. As a consequence the outward conductive energy flux can not be considered as a real energy loss, independent of the convective energy loss (Noble and Scarf, 1963; Kopp, 1968).

Nevertheless a fair estimate of the energy flux necessary to maintain the undisturbed corona is $F_m = 10^5$ erg/cm$^2$ sec.

The transition layer between the chromosphere and the corona is a thin layer with a thickness of the order of $10^3$ km in which the temperature increases by a factor 100. The importance of heat conduction in the transition layer is confirmed by temperature models derived from XUV spectral line observations (Athay 1966; Dupree and Goldberg, 1967; Burton *et al.*, 1969).

It is found that the temperature structure between $T = 10^5$ K and $T = 10^6$ to $2 \times 10^6$ K can be fitted into the relation

$$F_c = 6 \times 10^{-7}\, T^{5/2}\, (\mathrm{d}T/\mathrm{d}R) = \text{constant}, \tag{1}$$

which means that the greater part of the transition region is characterized by a constant conductive energy flux. It turns out that the energy loss from the corona by downward conduction is a few times larger than the total losses outward so that a flux of about $F_m = 5 \times 10^5$ erg/cm$^2$ sec must reach the level of maximum corona temperature.

The general picture is that this flux of energy is in the form of compression waves. The upper convective layers and the transition layers between the unstable and stable solar atmosphere, both in highly turbulent motion, are the likely source regions of the mechanical radiation i.e., the waves.

In a magnetic plasma in a gravitational field three fundamental wave modes appear which can be distinguished by the way the divergence and the curl of the velocity field are propagated (Lighthill, 1967), namely sound (acoustic) waves, internal gravity waves, and Alfvén waves. Sound waves can only propagate if the frequency $\omega > \omega_1 = \gamma g/2c$, where $c$ is the velocity of sound and $g$ is the acceleration of gravity. Gravity waves can only travel upwards if $\omega < \omega_2 = (g/H)^{1/2} \sim g/c$, where $H$ is the density scale height.

Under the combined action of non-parallel gravitational and magnetic fields mode coupling between gravity waves and Alfvén waves is possible. In this case all frequencies might be transmitted. This interesting possibility deserves more attention since it could well account for an effective energy transport in the magnetic parts of the solar corona. Outside the magnetic regions only sound waves and gravity waves contribute to the energy transport.

Sound waves can be generated efficiently in the turbulent layers of the convection zone as well as in the layers of turbulent penetration, while gravity waves can only be generated in the stable layers of turbulent penetration. There are also problems of propagation of the gravity waves, like strong radiative dissipation in the upper photosphere (Souffrin, 1966), and the small transmissivity at the photosphere-chromosphere temperature minimum (Uchida, 1967).

For both types of waves we know that, if the damping is not too strong, the amplitude must increase with height, approximately as $v \sim \exp(z/2H)$, because of the decrease in density. This amplification gives rise to a non-linear steepening of the wave profile until eventually a system of shock waves with accompanying rarefaction waves is formed. The average height of shock formation is several hundreds of kilometers above the photosphere.

At the time this occurs irreversible processes like viscosity and thermal conduction in the shock transition cause a transfer of the ordered energy in the wave motions into disordered thermal motions, which is called dissipation of mechanical energy. In a quasi-steady atmosphere the entropy rise after each shock is balanced by non-adiabatic processes during the subsequent expansion, which may be considered as a relatively slow isentropic expansion and an isobaric cooling. The amount of heating is directly connected to the net increase in enthalpy after the medium has been expanded to its initial pre-shock conditions.

Ultimately all energy losses from the solar corona must be replenished by the dissipation of mechanical energy.

A few theoretical models of the solar corona and transition region have been made and are essentially based on the energy equation

$$q_{\mathrm{conv}} + q_{\mathrm{cond}} + q_{\mathrm{rad}} = q_h, \tag{2}$$

where, for spherical symmetry,

$$q_{\mathrm{conv}} = \frac{1}{R^2}\frac{\mathrm{d}}{\mathrm{d}R}\left\{R^2\left(\varrho u\left[\tfrac{1}{2}u^2 + h - \frac{GM_\odot}{R}\right]\right)\right\} \tag{3}$$

is the energy which is convected outwards with the velocity $u$ of the solar wind, $h$ is the specific enthalpy and $GM_\odot/R$ is the potential energy.

The second term in Equation (2) is the energy loss by thermal conduction

$$q_{\text{cond}} = \frac{1}{R^2}\frac{\mathrm{d}}{\mathrm{d}R}\left(R^2\kappa\frac{\mathrm{d}T}{\mathrm{d}R}\right), \tag{4}$$

where $\kappa = 6\times 10^{-7}\,T^{5/2}$ for a fully ionized plasma. The other terms are the radiative energy loss $q_{\text{rad}}$, which in general is about proportional to $\varrho^2$ and $q_h$ is the heating per gram of matter.

For weak shock waves

$$q_h = \frac{16}{3}\frac{\varrho c^2}{P(\gamma+1)^2}(M_s - 1)^3, \tag{5}$$

where $c$ is the local sound velocity, $P$ is the period of the waves and $M_s$ is the shock wave Mach number.

The average distribution of the density and the temperature in a steady shock-heated atmosphere can be found by solving the Equations (2), (3), (4) and (5) together with an expression for the shock wave Mach number $M_s$ as a function of the distance, for which mostly approximate expressions are used. The solar wind velocity $u$ has to be found from the equation of momentum

$$u\frac{\mathrm{d}u}{\mathrm{d}R} = -\frac{1}{\varrho}\frac{\mathrm{d}p}{\mathrm{d}R} - \frac{GM_\odot}{R^2}, \tag{6}$$

and the equation of continuity

$$-\frac{1}{R^2}\frac{\mathrm{d}}{\mathrm{d}R}(\varrho u R^2) = 0. \tag{7}$$

The general problem is very complicated since the net flow velocity or solar wind velocity $u$ is not a free parameter but should fit the so-called critical solar wind solutions which will be discussed in the next section.

In the inner corona the solutions become less complicated because the convective energy flux plays a minor part in the energy transport. Farther out heating and radiative losses are negligible contributions to the energy equation. In these layers the temperature distribution is determined by heat conduction and convection. Still farther out there is fully adiabatic expansion of the corona.

According to the way the energy is transported Kopp (1968) divides the outer solar atmosphere into several distinct regions, as shown in Figure 1. Region I is the chromosphere, where mechanical heating is balanced by radiative losses. Region II is the lower transition region which is heated by thermal conduction and cooled by radiative losses. In region III, the upper transition region, thermal conduction is the main mode of energy transport. In region IV, the inner corona, the temperature maximum is reached. Shock wave heating is very important. Heat is conducted both

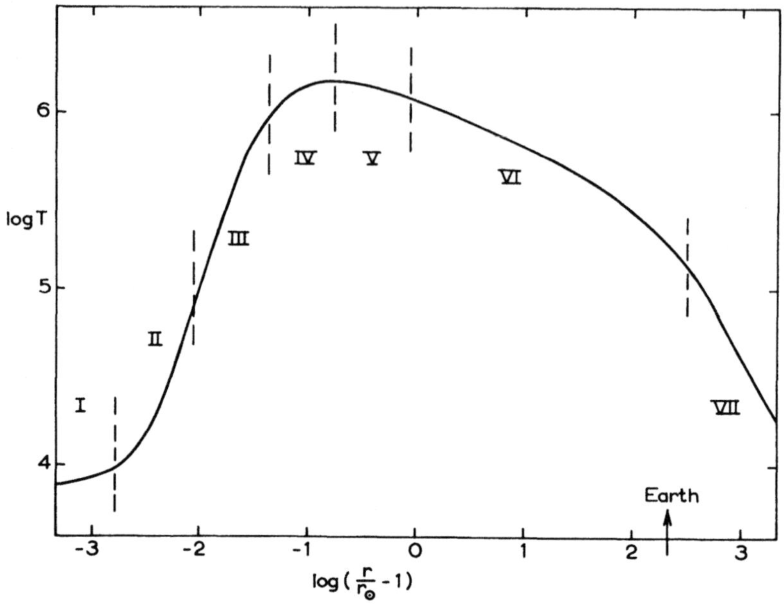


Fig. 1. Schematic division of the outer solar atmosphere according to energy-balance considerations (after Kopp).

into the outer corona and back into the chromosphere. In region V, the outer corona, hydrodynamic expansion becomes increasingly important. The upper boundary of region V is located at a level where the shock waves are exhausted. The temperature structure in region VI is entirely determined by the interplay of thermal conduction and hydrodynamic expansion, assuming there are no other heat sources. At large distances from the sun, region VII, the energy is convected outward only by adiabatic hydrodynamic expansion.

A discussion of models which has been recently constructed in the way outlined above is given by Kuperus (1969). Actually the problem of the heating of the corona is essentially to dissipate the right amount of energy at the right level.

Statements that enough energy is generated in the convection zone by turbulence or granulation to maintain the corona are misleading, since only a careful theoretical analysis can tell us under what circumstances the available energy is transferred to and dumped into the right levels.

## 3. Hydrodynamic Expansion of the Corona

The energy transport in the outer corona and in the interplanetary medium is almost completely achieved by the solar wind. Though discussion of the solar wind has frequently bypassed the problem of the heating of the corona it should be stressed that the hydrodynamic expansion of the outer solar atmosphere is a natural consequence of the heating of the corona to extremely high temperatures. The only excuse for a separate treatment is that in the lower layers (the inner corona) the solar wind plays a minor part and in the far outer layers the existence of a very high temperature is a sufficient basis for a discussion of the expansion.

A simple argument may be given as to why the corona is steadily expanding and why the atmosphere is not in hydrostatic equilibrium (Parker, 1963, 1965, 1969). If the corona were in hydrostatic equilibrium it could be easily demonstrated that one could obtain a vanishing pressure at infinity only if the temperature $T(r)$ should decrease more rapidly with the distance $R$ than $1/R$.

On the other hand, the corona as a fully ionized plasma has a very large coefficient of thermal conduction. In other words, for a constant energy flux one would obtain

$$\frac{1}{R^2}\frac{\mathrm{d}}{\mathrm{d}R}\left(R^2 T^{5/2}\frac{\mathrm{d}T}{\mathrm{d}R}\right) = 0,$$

using Equation (1). For large $R$, this expression yields $T \sim R^{-2/7}$.

This analysis was followed by Chapman (1957) and later by Unsöld (1960) to determine the temperature distribution in the corona. It is clear that a temperature determined only by heat conduction would decrease much too slowly for an atmosphere in hydrostatic equilibrium to fulfill the requirement of a vanishing pressure at infinity.

One therefore must conclude that the atmosphere is not in hydrostatic equilibrium but steadily expanding. There are several ways the atmosphere can expand. Most discussions are centered around the theory of stationary spherical expansion. We will show a few examples of nonspherical expansion at the end but since the physics is essentially the same we shall omit this complication for the moment.

The important equations are the equation of motion and the equation of continuity. From the Equations (6) and (7), after elimination of the density, Parker (1965) finds the solar wind equation

$$u\frac{\mathrm{d}u}{\mathrm{d}R}\left(1 - \frac{c_1^2}{u^2}\right) = -R^2\frac{\mathrm{d}}{\mathrm{d}R}\left(\frac{c_1^2}{R^2}\right) - \frac{w^2}{R^2}, \tag{8}$$

where $c_1 = 2kT/M$ is the thermal velocity and $w = GM_\odot/R_0$ is the gravitational escape velocity at the base level $R_0$ of the corona.

Equation (8) can be solved for the velocity as soon as the temperature is known as a function of the distance, which is only possible when the energy Equation (2) is solved simultaneously. For a few simple cases, like an isothermal atmosphere or an adiabatically expanding atmosphere, a solution is possible without using the energy equation.

Since the corona is an atmosphere which at its base is strongly bound to the sun, the right-hand term is negative at the base of the corona. Actually, the thermal energy at the base is an order of magnitude smaller than the gravitational energy. However, at great distances the right-hand term is positive if $T(R)$ decreases less rapidly than $1/R$. Hence, at a certain critical distance $R_c$ the right-hand term vanishes. Now since we are only interested in solutions with small $u_0$ we see that $(\mathrm{d}u/\mathrm{d}R)_{R=R_0}>0$. If $u(R_c)<c_1$, $(\mathrm{d}u/dR)_{R=R_c}=0$ and it will always be the case that $u<c_1$. This class of solutions is called the subsonic solutions or 'solar breeze'. The trouble with subsonic solutions is that they yield a finite pressure at infinity. Only if $u(R_c)=c_1$ will the velocity grow to supersonic values. This solution is a critical solution but it is the only solution that is consistent with the observations of the solar wind.

The distance of the critical point depends on the temperature of the corona. For an isothermal atmosphere it is given by

$$R_c = GM_{\odot}M/4kTR_{\odot}\,. \tag{9}$$

$R_c$ is a few solar radii for coronal temperatures.

Whether the corona is expanding supersonically or subsonically depends essentially on the energy transport in the lower corona. It can be shown that for a fixed temperature at the base there is a maximum density at the base of the corona, $N_0(\mathrm{max})$ below which supersonic expansion is possible under the assumption that the corona is heated by thermal conduction only. The explanation is that when the density at the base is increased the convective energy flux is increased, while the conductive energy flux is independent of the density. However, the convective flux can only increase at the expense of the conductive flux thereby causing a change in the temperature gradient. The critical density $N_0(\mathrm{max})$ strongly depends on the temperature. For the quiet corona with $T\approx 1.5\times 10^6$ K, $N_0(\mathrm{max})\approx 10^9$ cm$^{-3}$. For the active corona $N_0(\mathrm{max})$ drops to a value of $10^7$ cm$^{-3}$. However, it seems that above active regions the corona still expands supersonically. This would mean that considerable heating by dissipation of wave energy would take place at distances much further out than the so-called base of the corona. The idea that extended heating must take place was put forward by Noble and Scarf (1963).

## 4. The Influence of Magnetic Fields on the Structure of the Corona

So far we have discussed the dynamics and structure of the corona as if there the structure were homogeneous or as if the atmosphere were spherically symmetric. White-light photographs clearly demonstrate that there are large deviations from spherical symmetry which are believed to be caused by the magnetically inhomogeneous structure in the corona. The effects of the magnetic field on the heating and the dynamics are so numerous that at this moment a theory that gives full credit to the magnetic field for even the simplest field configuration does not exist.

However many discussions have appeared on certain aspects of the field and, in a sense, these investigations can be of great use in particular layers of the corona under particular circumstances. The main influence of a magnetic field is that it adds a new preferential direction to the plasma besides the direction of gravity or the direction of wave or heat propagation. As we have learned, it is likely that more energy is generated in waves in the magnetic regions and therefore we expect a different atmosphere above the magnetic regions. Whether the corona above magnetic regions is hotter and/or more extended depends essentially on the topology of the field, as will be demonstrated in the example at the end of this section.

The way in which the energy is transported in this case is not a simple compression-type wave, since the waves are strongly modified by the magnetic field. Fast, slow, and Alfvén modes may occur depending on the local circumstances and on the way they can be excited. It is difficult to predict how much energy will be fed into these wave modes.

In the photosphere outside active regions the magnetic fields are small and the energy is mainly transported by sound waves and internal gravity waves. At greater heights the Alfvén velocity is certainly larger than the sound velocity and part of the mechanical energy flux may be picked up by the Alfvén mode. The study of the non-linear coupling of magnetohydrodynamic wave modes is not yet far enough evolved to make even an approximation of the energy flux in Alfvén waves, although Lighthill (1967) has done some interesting work. Alfvén waves are probably weakly damped, and this offers the possibility of carrying a substantial amount of energy to the far outer layers for which, as we have seen, additional heating is needed. Barnes (1968) has made a study of the heating of the far outer corona and interplanetary medium by magnetohydrodynamic waves.

An important consequence of the presence of a magnetic field which is relatively easy to incorporate is the strong channelling of heat flow along the magnetic field lines. The coefficient of thermal conduction is strongly reduced in the direction at right angles to the magnetic field. This means that the heating and structure of the layers where thermal conduction is the main mode of energy transfer (transition region and outer corona) are entirely determined by the geometry of the magnetic field. This may well be the reason why the visible structures in the outer corona appear to reflect the magnetic field geometry.

The influence of a magnetic field on the structure of the transition layer has been discussed by Kopp and Kuperus (1968). The channelling of heat along the magnetic field results in a concentration of the heat in the regions where the magnetic field is most concentrated. Because of the channelling of the conducted heat flux the temperature gradient in the transition region is much larger in the regions where the field is concentrated (the spicule regions) than it is in the so-called interspicule regions. Figure 2 shows schematically a temperature structure of the transition region and inner corona for a given magnetic field configuration.

It has been shown by Dubov (1970) that this model of the transition layer, together with an inhomogeneous model of the underlying layers (Dubov, 1968) is able to explain the observed emission in the optical, radio and UV spectral range. One would conclude that in magnetic regions more energy is available to heat the outer layers and that the heating mechanism in large parts of the corona occurs along the field lines.

An interesting circumstance arises when young magnetic regions appear where the magnetic field emerges in the form of ropes or flux tubes and does not extend very far into the corona. In this case the enhanced heating serves primarily to heat the coronal center of activity to higher temperatures than the surrounding atmosphere.

This may go on until the pressure has been built up to such a high value that the whole field configuration is blown up and eventually a streamer is formed (Pneuman, 1968a, b). The initially closed field may also be opened and transformed into a streamer after the occurrence of a solar flare (Kuperus and Tandberg-Hanssen, 1967).

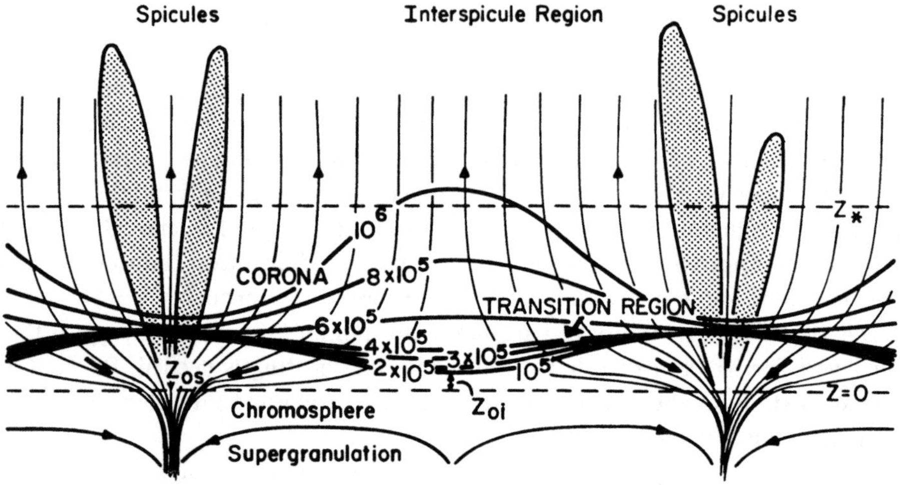


Fig. 2. A schematic cross-section of the region from the upper chromosphere to the lower corona showing magnetic field lines, isotherms, and directions of the observed motions.

As soon as the initially closed field is opened, the solar wind can take over a large part of the energy transport. It is clear that the temperature and density structure in a magnetically closed region is completely different from the structure in open streamers because of the regulating effect of the solar wind. For example, one cannot increase the temperature above a certain value, about 2 to $3\times10^6$ K, because at higher temperatures the convective flux increases tremendously. It is like boiling water. Putting more heat in the water does not increase the temperature. All the extra heat goes into evaporation of the water. Therefore the very hot coronal condensations are either of extremely short duration or they are confined in closed magnetic loops.

The structure and the energy transport in a non-radially extended streamer has been investigated by Pneuman and Kopp (1970). For an open streamer configuration they find that the maximum density at the base $N_0$(max) for which there is supersonic expansion is lower in the streamer than for radial expansion. Apparently the conductive flux in the streamer is suppressed more than the convective flux, resulting in a predominance of the convective flux. This leads to a lower base density below which supersonic expansion occurs.

However, since subsonic expansion in magnetically aligned streamers has not been observed, Pneuman and Kopp conclude that the dense parts that would lead to subsonic expansion probably occupy a smaller part of the atmosphere at high layers than at low layers. Another alternative is that open streamers only occur above a certain height. Below this level everything is confined in closed field lines (domes, helmets). They also mention the possibility that extended heating other than thermal conduction may be responsible for the observed supersonic expansion. A comparison between the density, temperature, and solar wind velocity in and outside an open streamer as found by Pneuman and Kopp is given in Figure 3. The streamer was characterized by a throat at some distance. It was found that the temperature gradient in the layers below the throat is smaller than in the layers above.

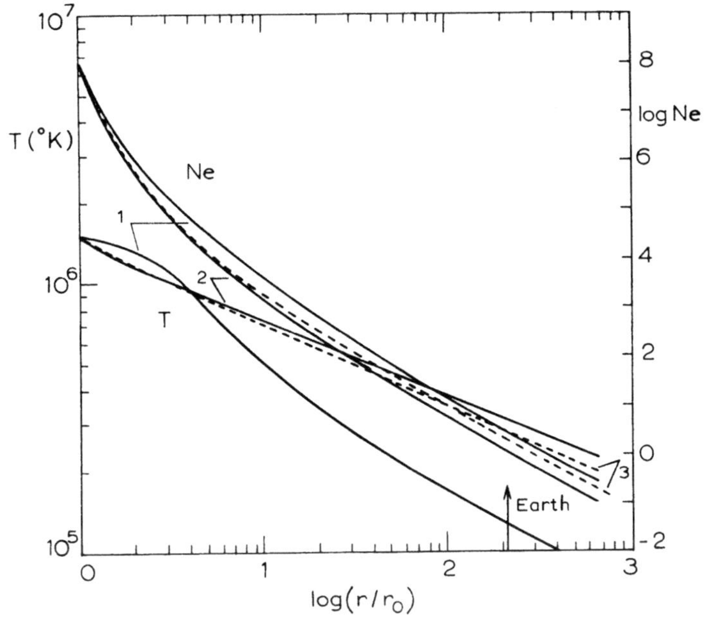


Fig. 3. Radial variation of streamer (1) and interstreamer (2) temperature and density as compared with the temperature and density distribution in a radially expanding corona (3) (after Pneuman and Kopp, 1970).

These investigations strongly support the conclusion that the presence of a magnetic field which channels the heat flow is a primary cause for the inhomogeneous structure of the solar corona.

## References


Altschuler, M. D. and Newkirk, G.: 1969, *Solar Phys.* **9**, 131.
Athay, R. G.: 1966, *Astrophys. J.* **145**, 784.
Barnes, A.: 1968, *Astrophys. J.* **154**, 751.
Billings, D. E.: 1966, *A Guide to the Solar Corona*, Academic Press, New York.
Burton, W. M., Jordan, C., Ridgeley, A., and Wilson, R.: 1969, H.A.O. Conference on the Chromosphere-Corona Transition Region, held in Boulder, Colorado, p. 35.
Chapman, S.: 1957, *Smithson. Contrib. Astrophys.* **2**, 1.
Dubov, E. E.: 1968, in K. O. Kiepenheuer (ed.), Structure and Development of Solar Active Regions, *IAU Symp.* **35**, 255.
Dubov, E. E.: 1970, *Izv. Krymsk. Astrophys. Observ.* **45**.
Dupree, A. K. and Goldberg, L.: 1967, *Solar Phys.* **1**, 229.
Kopp, R. A.: 1968, Thesis Harvard University, Scientific Report No. 4, AFCRL-68-0312.

Kopp, R. A. and Kuperus, M.: 1968, *Solar Phys.* **4**, 212.
Kuperus, M.: 1969, *Space Sci. Rev.* **9**, 713.
Kuperus, M. and Tandberg-Hanssen, E.: 1967, *Solar Phys.* **2**, 39.
Lighthill, M. J.: 1967, in R. N. Thomas (ed.), 'Aerodynamic Phenomena in Stellar Atmospheres', *IAU Symp.* **28**, 429.
Neugebauer, M. and Snyder, C. W.: 1966, in R. J. Mackin and M. Neugebauer (eds.), *The Solar Wind*, Jet Propulsion Lab. and Cal. Inst. of Technology, Pasadena, Calif., p. 3.
Newkirk, G.: 1967, *Ann. Rev. Astron. Astrophys.* **5**, 213.
Noble, L. H. and Scarf, F. L.: 1963, *Astrophys. J.* **138**, 1169.
Parker, E. N.: 1963, *Interplanetary Dynamical Processes*, Interscience, New York.
Parker, E. N.: 1965, *Space Sci. Rev.* **4**, 666.
Parker, E. N.: 1969, *Space Sci. Rev.* **9**, 325.
Pneuman, G. W.: 1968a, *Solar Phys.* **4**, 578.
Pneuman, G. W.: 1968b, *Solar Phys.* **5**, 595.
Pneuman, G. W. and Kopp, R. A.: 1970, *Solar Phys.* **13**, 176.
Shklovskiy, I. S.: 1965, *Physics of the Solar Corona*, Pergamon, London.
Souffrin, P.: 1966, *Ann. Astrophys.* **29**, 55.
Uchida, Y.: 1967, in R. N. Thomas (ed.), 'Aerodynamic Phenomena in Stellar Atmospheres', *IAU Symp.* **28**, 456.
Unsöld, A.: 1960, *Z. Astrophys.* **50**, 57.
Van de Hulst, H. C.: 1953, in G. P. Kuiper (ed.), *The Sun*, Chicago Univ. Press, Chicago, p. 207.

# LARGE-SCALE MAGNETIC FIELDS AND ACTIVITY PATTERNS ON THE SUN

V. BUMBA

*Astronomical Institute of the Czechoslovak Academy of Sciences, Ondřejov*

**Abstract.** The recent results concerning the large-scale magnetic field and solar activity distribution are summarized. The static characteristics for three types of such large-scale features are given: (a) *Giant regular structures* (super-supergranules) are probably the largest elements in the hierarchy of solar convective cells, which can be defined in the solar atmosphere for example, by magnetic field patterns, faculae, calcium flocculi, and the organization of active regions. (b) *Active longitudes* are the longitude sectors with grouped activity and concentrated magnetic fields. The grouping lasts in the same sector for many successive rotations. These active longitudes may have different rates of shift in heliographic longitude per rotation in a given system of coordinates. They are demonstrated not only on magnetic synoptic charts arranged by latitude zones but may also be followed on Zürich Heliographic Maps of the Photosphere for at least four cycles. This means that the organization of activity and fields in streams or active longitudes does not change during several tens of years and proceeds in the same way following the same rules of development. (c) *Active latitudes* may also play a role in the formation of large-scale patterns.

The dynamics of large-scale features is presented. The active longitudes develop as consecutive *complexes of activity* which are separate impulses of activity, being usually in opposite phase if two longitudes are active. Giant regular structures develop analogously to supergranules or granules if the background magnetic field in the active longitude reaches a certain level. A possible theoretical explanation of the formation of large-scale patterns, including the 'influence of the planets' is discussed.

The sun as a star and the manifestation of longitudinal and latitudinal distributions of the solar magnetic field in the interplanetary magnetic field is considered. Problems of long-term forecasts of important solar events on the basis of recurrences in the large-scale magnetic field and activity patterns are shown.

## 1. Introduction

Although the study of the large-scale distribution of solar activity over the solar disc is practically as old as the investigation of solar activity itself, only systematic measurements of weak solar magnetic fields on the whole disc, like those made at the Mount Wilson Observatory, have enabled observers to see the large-scale magnetic field patterns. More than a ten year period of observations of this basic characteristic of solar activity is now available.

During recent years it has been demonstrated more and more that the development of a single active region and particularly that of complicated active regions may be better understood if it is studied from the point of view of its relation to the distribution of magnetic fields at all scales, from the smallest to the largest. At the same time more evidence has been found on the connection of solar activity as a whole with the existence of the large-scale features in the solar atmosphere. Moreover, because they are long-lived and often regularly spaced over the solar surface, they may be used for long-term forecasts of solar activity, as soon as we succeed in identifying the time and the location of the beginning of their formation.

Because of their size one would expect that active regions extend into very deep

*Dyer (ed.), Solar Terrestrial Physics / 1970: Part I,* 21–37. *All Rights Reserved.*

layers of the sun. If so, they are very important for the solution of certain theoretical problems, such as the dynamics of convection, the intensification and dissipation of magnetic fields, the maintenance of differential rotation, the problem of the solar activity cycle, etc. Because of their size they also significantly influence the distribution of magnetic fields, not only in the outermost layers of the solar atmosphere, but also in interplanetary space, where they constitute one of the most pronounced characteristics.

## 2. Typical Large-Scale Features (Statics)

If we summarize all facts about the large-scale patterns of solar magnetic fields and activity yet known, we may conclude that there are several basic factors influencing the morphology of magnetic fields and the distribution of activity in the solar atmosphere, while the physical meaning of each of these factors evidently need not be the same:

(a) '*Giant regular structures*' are most probably the largest elements of a system of convective motions formed in plasma and magnetic fields of solar atmospheric layers (Bumba *et al.*, 1964; Simon and Weiss, 1968a). There is a hierarchy of convective cells of different-scale granulation ($1–2 \times 10^8$ cm), supergranulation ($3–4 \times 10^9$ cm) and 'giant regular structures' or 'super-supergranules' ($3–4 \times 10^{10}$ cm) which seems to represent in a way the influence of the 'quiet sun' on the field distribution. This hierarchy of convective cells is one of the most important agents in the concentration and dissipation of magnetic fields (Bumba, 1967).

The 'giant regular structures' are observable in the photosphere and chromosphere as magnetic field patterns, faculae, the calcium emission flocculi, and the organization of individual active regions. Their shape resembles a rhombus with the longer sides parallel to the equator and about 30°–35° long, slightly longer than the other sides, which are at an angle to the equator and about 25°–30° long. The individual structures form a regular network best seen in years of moderate solar activity. At times of high solar activity the background magnetic field forms one body with regular or semiregular patterns in between and it is difficult then to follow the shapes of the giant regular structures. The active regions occur mostly in the crossings of these giant cells. During years of minimum activity the background field is too weak and disappears too soon for the giant structures to be formed (Bumba, 1970). The regularity of the magnetic field distribution of this type, with maxima spaced about 35° apart in longitude, was checked with the aid of an autocorrelation analysis of seven magnetic synoptic charts (Bumba *et al.*, 1969). The strong maximum at about 60° found by this analysis may be interpreted as due to a network of 'giant structures' which is, as we shall see, very closely related to another large-scale feature, the 'active longitudes'.

(b) '*Active longitudes*' are the longitude sectors in which activity is grouped together, usually in a number of active regions. The question of whether 'active longitudes' are real, which is about eighty years old, was revived by Warwick (1965) and Levitskij (1967). In particular Dodson-Prince and Hedeman (1968) reopened the question with a demonstration of the concentration of the greatest part of solar activity during 1962–1967 in two Carrington heliographic longitude sectors almost 180° apart. Švestka (1968a, b), Švestka and Simon (1969) looked for 'dangerous longitudes' on the sun in relation with proton-flare forecasts, and showed that proton-flare activity was concentrated in certain heliographic longitudes from studying their shift during a period of about one solar cycle. Vitinskij (1969a, b) investigated by his method of isolines the manifestation of sunspot and flare activity in 'active longitudes' for the years 1957–1962 and could show that the rate of concentration in 'active longitudes' is the highest for the more important active formations.

Using the synoptic charts of weak fields from the 'Atlas of Solar Magnetic Fields' observed at the Mount Wilson Observatory (Howard *et al.*, 1967) Bumba and Howard (1969a, b) not only showed the reality of 'active longitudes' but also interpreted them as recurrences in the solar large-scale magnetic field distribution. The features of the background field of one polarity may be followed on subsequent synoptic charts usually for at least 10 rotations so that if, for example, we arrange the equatorial strips ($\pm 20°$) of such charts in chronological order we see these regions form long rows. Of course, these patterns depend on the level of activity, but very often the individual rows constitute a nearly continuous stream which may be followed for several years. In periods of greater activity two main streams, most probably identical with two 'active longitudes', are seen. The secondary stream is distant from the first stream by about 200° to the East or 160° to the West. This slight asymmetry from the 180° distribution seems to be significant. When the background-field pattern is strong, the streams are not easily seen; the rows, however, are very striking in appearance at this time. During periods of moderate and low activity the streams or at least the main one is quite evident, but the level of activity may be so low that the strong regularity of the rows is no longer seen. It was shown that not only the development of spot groups follows the streams, but also flare activity, especially proton-flare activity, is concentrated in the longitudinal sectors of the main streams (Bumba and Obridko, 1969).

The inclination of rows and streams in given heliographic coordinates (usually Carrington coordinates are used) may be different, which means that they have different longitudinal drifts. In the magnetic background-field main inclinations corresponding to rotational values of large-scale regularities of about 27.0 and 28.0–29.0 days were found, although these were not the only value which could be seen. The fine structure of rows and streams, the intensification of subsequent individual rows and streams in a regular succession or, in other words, a certain kind of periodicity in the development of activity in individual streams, all suggest the complexity of the phenomenon.

So far we have discussed the results obtained from the background magnetic field study. Knowing such regularities and using the fact that photospheric faculae are seen only in that region of the supergranular network where the intensity of the concentrated magnetic field reaches a certain value, we may take for our investigation of long-lived large-scale regularities in activity and magnetic field distribution and also the course of 'active longitudes' the Zürich 'Heliographic Maps of the Photosphere', which comprise synoptic charts published since 1937. In the publications of the Eidg. Sternwarte Zürich (Wolfer, 1897) it is even possible to find and use for our study photospheric synoptic charts for the period of the years 1887 till 1898. In this way we can extend our observational material to more than forty years in which the last ten years coincide with those covered by the magnetic field measurements. Using the same method of evaluation as that applied to the magnetic synoptic charts, i.e. the separation of the equatorial ($\pm 20°$) and higher latitude ($\pm 20°$ to 40°) zones and the organization of such strips of maps in chronological order, (Figure 1, Figure 2, Figure 3) we note that (1) there is practically no difference between the magnetic field and facular streams (except the alternation of the polarity of magnetic field rows); (2) there is practically no difference between the character of the distribution of facular streams in the four individual cycles of solar activity for which we have observational material. On the contrary, the fine structure of streams seems often to be more manifest; the variety of inclinations to be better recognized, and the intensification of individual streams in a certain succession or the streams formed from the maxima of secondary smaller streams with different inclination to be better shown.

For the investigation of the fine structure of the distribution of inclinations of individual streams, Bumba and Letfus (1971) have made a preliminary comparison (Figure 4) of a histogram of such inclinations, graphically estimated from the first part (1887–1898) of the synoptic map material from both equatorial and high latitude strips with the power spectrum resulting from an autocorrelation analysis of daily sunspot areas published in the 'Greenwich Heliographic Results' from 1883–1959. These preliminary results already indicate a relatively good coincidence of the fine structure of the distribution of individual secondary peaks of rotational periods around the main broad 27-day maxima in both graphs. The physical meaning of this fact seems to be obvious: the organization of solar active regions (represented by the facular fields) in various streams or active longitudes has not changed during the time interval 1883–1959 (1970) and has proceeded in the same way following the same rules for practically a whole century. We shall discuss the physical meaning of streams, with different intensity and inclination, in one of the later sections.

(c) *Active latitudes.* It is commonly accepted to speak of Spörer's law and its influence on the distribution of activity in heliographic latitude. Just recently it was demonstrated (Gnevyshev, 1963; 1967; Antalová and Gnevyshev, 1965) that, instead of a smooth change of heliographic latitude of solar activity phenomena with the progress of the cycle, two distinct maxima in two separate latitude zones occur during the cycle: the first, coinciding with the main maximum of cycle activity, is tied to relatively high latitudes; the second appears about three years later and is closely related to low latitudes. This behaviour in the development of solar activity, which may be observed in the variation of spot number and area, prominence area, green corona intensity, and the radio flux, and which is also well known from the frequency distribution of large, especially proton-flares during the cycle (Švestka, 1966), may

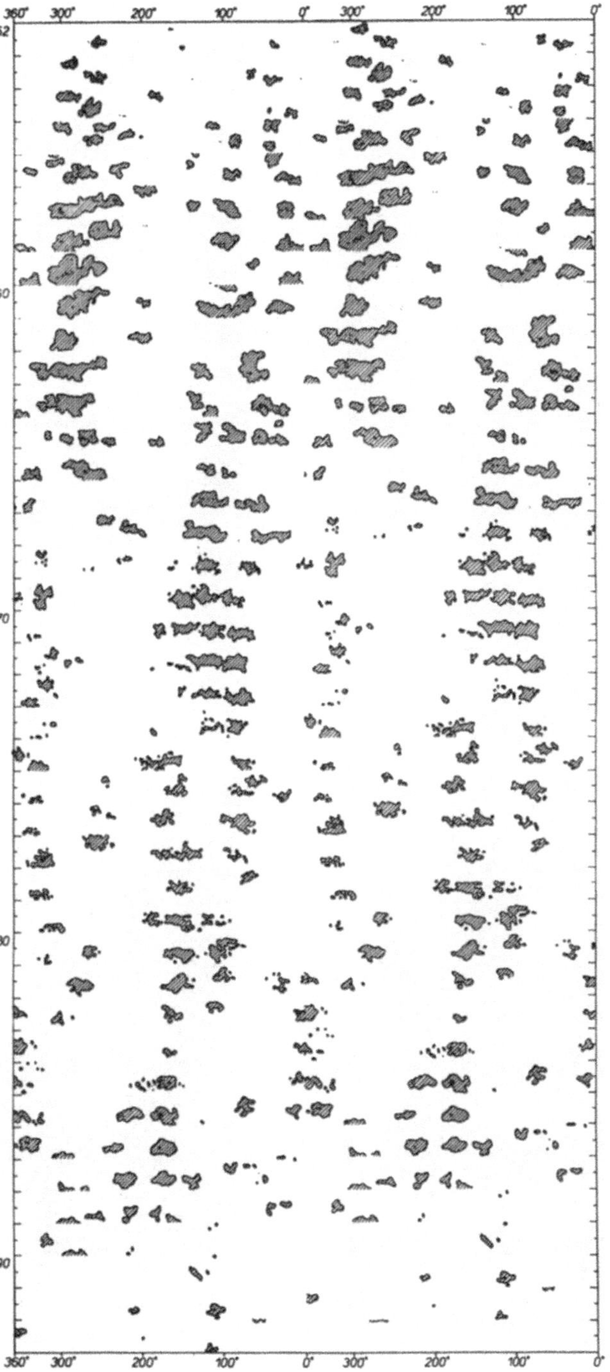


Fig. 1. Streams of facular fields (active longitudes) from the end of the last century (rot. Nos. 352–394) during a time of relatively low activity. The synoptic charts are drawn twice.

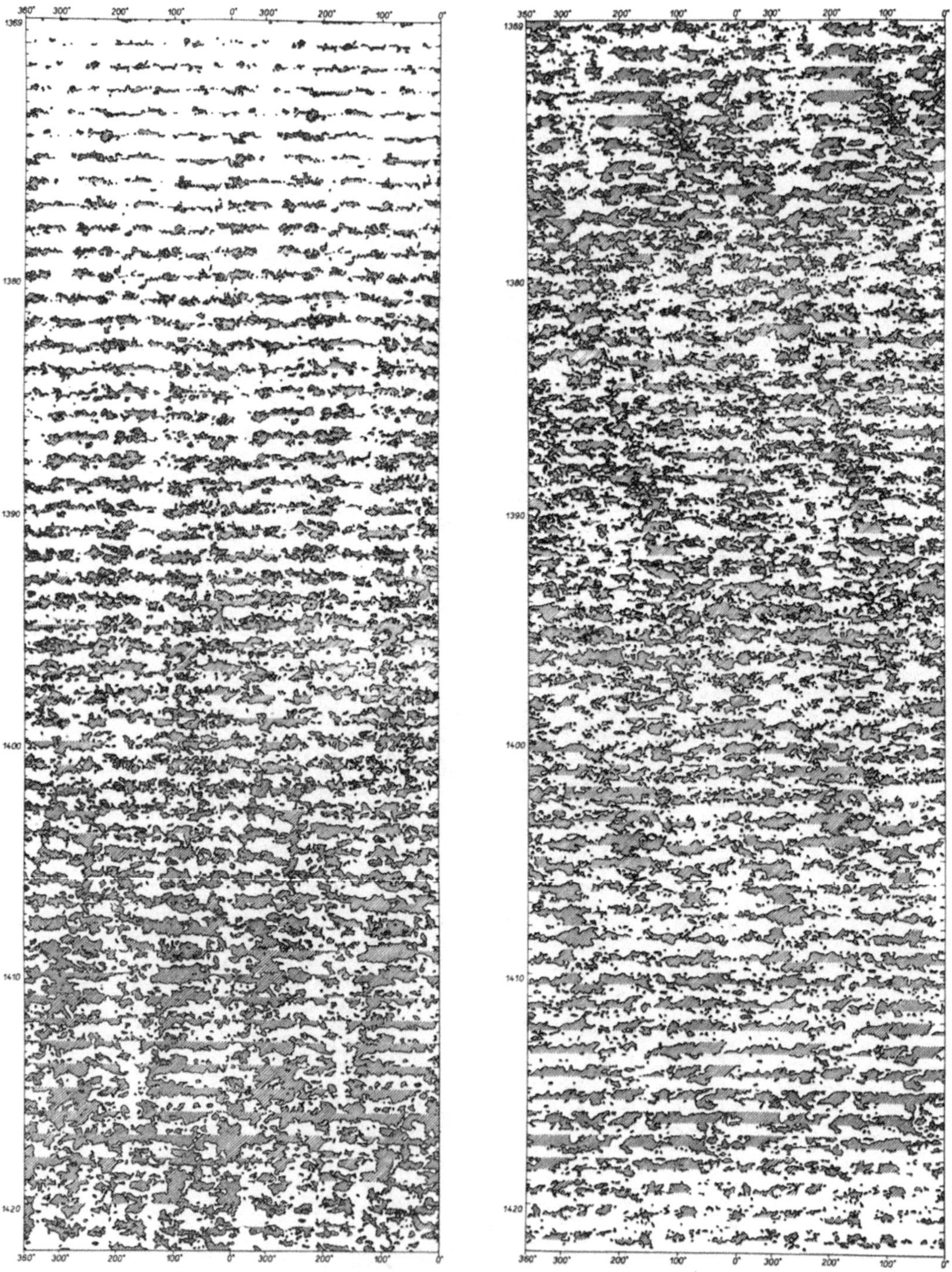

Fig. 2. Streams of facular fields (active longitudes) for rotations Nos. 1369–1420 i.e. in a period of growing activity. Left part: the equatorial zones (±20°). Right part: the higher latitude zones (±20°, to ±40°). In both pictures the synoptic charts are drawn twice.

Fig. 3. Streams of facular fields (active longitudes) for the rotations Nos. 484–513 i.e. in a period of higher activity. The synoptic charts are drawn twice.

be also traced in Sun-Earth relations (Gnevyshev, 1967) and even in the modulation of cosmic ray intensity (Balusubrahmanyan and Venkatesan, 1970).

From the study of synoptic maps the concentration of a polarity region in certain heliographic latitudes, at least during particular stages of activity development, is known (Bumba and Howard, 1965a; Bumba and Růžičková-Topolová, 1969). Also

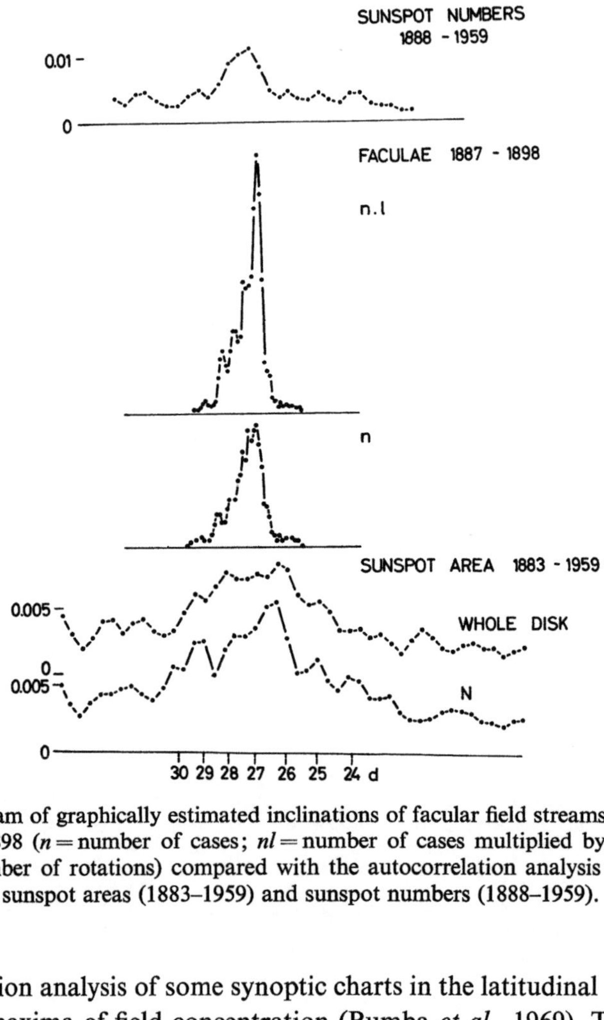


Fig. 4. A histogram of graphically estimated inclinations of facular field streams from the synoptic charts for 1887–1898 ($n$ = number of cases; $nl$ = number of cases multiplied by the length of the stream in the number of rotations) compared with the autocorrelation analysis of the maxima of sunspot areas (1883–1959) and sunspot numbers (1888–1959).

the autocorrelation analysis of some synoptic charts in the latitudinal direction shows two separated maxima of field concentration (Bumba *et al.*, 1969). The study of the magnetic field, green corona and filament distribution in high solar latitudes during 1959–1961 (Bumba and Růžičková-Topolová, 1969) did not show any shift of magnetic fields and corona emission beyond the heliographic latitude $\varphi = 65°$ during the whole time interval investigated.

All these results evoke not only many questions connected with the transport of magnetic fields from activity zones and with the meridional circulation in the solar atmosphere; they may also be connected on one side with the existence of 'large regular structures' ('super-supergranules') and on the other side with the possible existence of preferential 'active latitudes' for the appearance of solar activity.

The situation is still further complicated by the results of some magnetic field observations (Bumba, 1967) and measurements of solar rotation (Livingston, 1969). These investigations, made with the aid of a photoelectric magnetograph, have shown that the shape of magnetic background field patterns and curves of solar rotation as a function of solar latitude may indicate that the velocity of solar rotation changes suddenly just at the high-latitude boundary of the main background-field body (or the zone of activity) where the magnetic fields begin to be less crowded. In other words, this could mean that the presence of a magnetic field body in the zones of activity could force the solar photosphere to rotate at least partly as a rigid body.

The problem of the latitudinal structure of large-scale patterns of field and activity seems to be at this moment one of the least investigated ones.

## 3. Large-Scale Features and Solar Activity (Dynamics of Their Development)

In the previous section we have tried to characterize the basic static elements of the large-scale magnetic and activity patterns. But to understand better their mutual relationships we have to discuss first of all another type of large-scale feature which has a more dynamical character:

(a) A *complex of activity* is a concentration of activity in a separate impulse over a limited range of heliographic longitude – probably a kind of a basic unit in which the activity is realized (Bumba and Howard, 1965a; Bumba, 1967). Complexes usually develop in a longitudinal range (active longitude) over which one only observes during several rotations old magnetic fields of dissipating active regions. In complexes the activity behaves as in other phenomena: it rises rapidly to its maximum and is followed by a slow decline of activity. The area occupied by fields expands in both latitude and longitude. The expansion in the north-south direction is limited by the heliographic latitude of the edge of the main background-field body. The rate of shift of the magnetic field boundary in longitude is different for the leading and following parts but each value of the velocity is practically the same for all the complexes studied. In general, the first sunspot groups to form are the largest of the complex. The larger the complex, the larger its first spot groups and the longer the lifetime of the complex (10–12 rotations). From large complexes (in which about 15–20 active regions are formed consecutively) the 'unipolar magnetic regions' develop, and the very complex patterns of the background field.

Complexes of activity are easily observable during years of moderate solar activity. In years of greater activity it is very difficult to distinguish individual complexes, although the activity changes and the appearance of synoptic charts show that solar activity may still occur through the individual complexes.

(b) *Development of 'active longitudes'*. If we compare the position of individual complexes of activity with the large-scale distribution of magnetic fields on a sequence of synoptic charts, we see that there are streams which are closely related to the complexes, and (vice versa) there are complexes which produce the magnetic fields forming the streams. This means that the active longitudes develop as a series of successive complexes of activity. This fact may be followed very well if only one or two active longitudes on the sun are present. In the case of one longitude practically all activity is concentrated in it and its time variation has all the characteristics of development of a complex of activity. The same is true if two main longitudes exist, with the only difference that the development of activity in each longitude is very often in an opposite phase.

Also the periods of rotation of the west and east boundaries of the complex are in striking coincidence with the two main inclinations of streams: 27.0 and 28.0–29.0 days (Bumba *et al.*, 1968; Sýkora, 1969).

(c) *Development of 'giant regular structures'*. 'Giant regular structures' are observable only in such a region of the solar surface where the activity and therefore the supply of new magnetic flux appears over at least several rotations, this means only in such active longitudes in which still new complexes of activity develop. They are best seen during those time intervals when only one or two 'active longitudes' occur so widely separated that the redistribution of magnetic fields in the convective elements is not influenced by other activity in the neighbourhood. At the same time the level of the background magnetic field intensity should just reach the value necessary for the best visibility of the giant regular structures. This depends, of course, on the stage of activity in the complex (or active longitude) and consequently, because of the phase shift of activity development in various longitudes, it takes place during a given time interval usually in only one of the longitudes.

The "giant regular structures" (Bumba, 1970) (see Figure 5) develop analogously to supergranules or granules, which means that one structure is gradually transformed into another one of the same scale through the gradual redistribution of supergranules and individual active regions on their peripheries. The life-time, or better to say, the visibility of the 'super-supergranules' is about five, sometimes even more rotations, of which about two or three rotations are needed for the structure to be formed and about three or four rotations to fade away or to be transformed into another structure. The active regions of the complex develop first in the crossings of these giant features. The later, smaller active regions proceed in their development like the emission of flocculi during the formation of a single active region (Bumba and Howard, 1965b) following the periphery of neighbouring supergranules, only the scale is one order greater: these secondary active regions usually go again from the crossings in both directions in the space between giant structures.

(d) *Concerning the theoretical explanation of the formation of large-scale features.* If we try to summarize briefly the dynamics of the development of large-scale features in the solar magnetic field and activity distribution we see that there exists a hierarchical system of convective motions which seems to rearrange the lines of force of magnetic fields supplied in certain quanta into the upper layers of the solar atmosphere in active longitudes where the activity changes periodically through the development of complexes of activity. The role of possible active latitudes does not yet seem to be

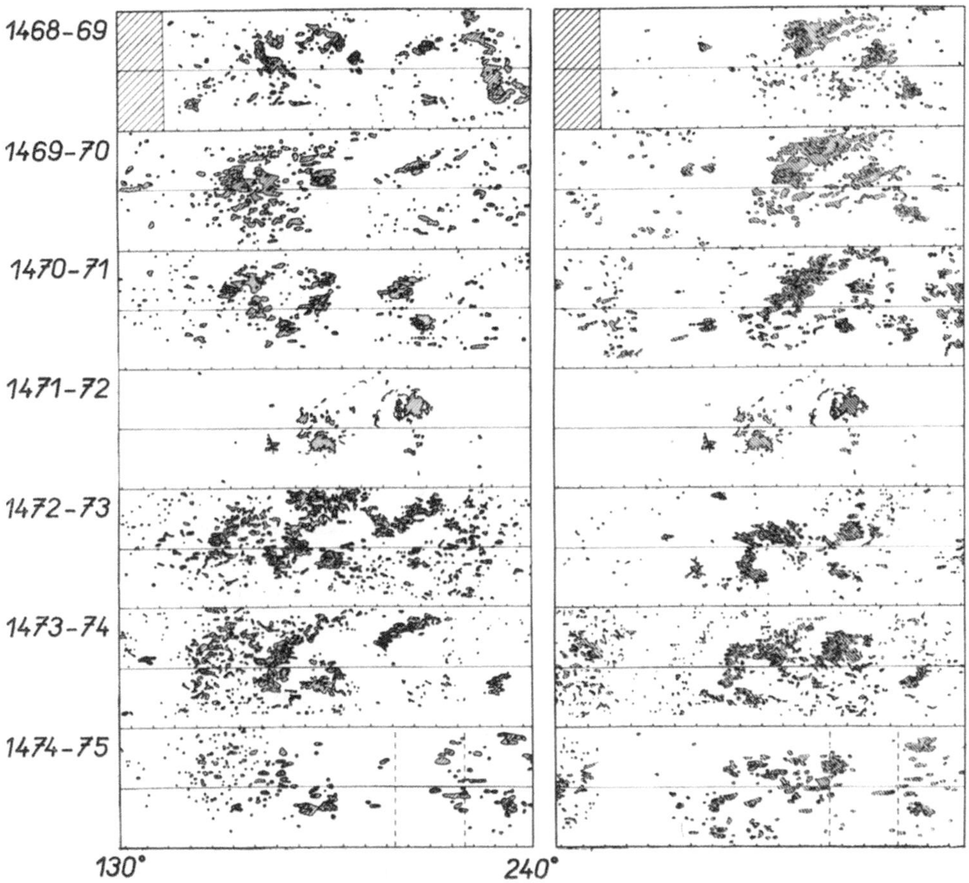


Fig. 5. The development of 'giant regular structures' in the plus polarity field in 1963. The polarities are separated in the figure: the left side represents the plus polarity synoptic charts, the right side the minus polarity maps.

fully understood: it may for example enlarge the influence of active longitudes in the given latitude.

It seems clear that for the theoretical explanation of the processes described earlier it is necessary to investigate on one side the giant regular structures (or the 'super-supergranules') as possible convective cells, formed probably near the base of the convective zone, and on the other side the active longitudes in which submerged sources of activity may show themselves (but because of their rich fine structure and different inclinations other agents may also be expected to exert a strong influence on their formation). The morphology of complexes of activity is probably defined by all these effects, as soon as a certain portion of magnetic fields and activity accompanying them develops in an active longitude.

A simplified model of the convective zone which explains not only the generation of both photospheric granules and supergranules but also suggests the existence of giant cells was presented recently by Simon and Weiss (1968a, b). Instead of assuming that the dimension of the cells is limited to the density scale-height this model proposes an extension of cellular motion over several scale-heights. The horizontal component of such a motion sweeps magnetic flux aside and concentrates it on the edges of the cell. This process stops when the energy of the concentrated field is comparable with the kinetic energy of the horizontal motion (Simon and Weiss, 1968; Bumba, 1967).

With regard to the magnetic streams the dynamics of their development is known (Bumba and Howard, 1965a): they are the results of the expansion and weakening of old active regions and their interaction with neighbouring fields, and of the continuing development of magnetic fields of new regions within the pattern. It is also evident that new regions (Bumba and Howard, 1965b) are formed in or immediately adjacent to expanding weak old regions. Furthermore, it is evident that the newly formed regions have not appeared from an immediate reorganization of old magnetic fields, because old features show little or no changes during the appearance of new regions. We have mentioned that they develop in crossings of giant cells, in which the very old fields were rearranged. But it is still difficult to decide what is the principal process developing these new formations supplying photosphere with a new flux of magnetic fields. The theorists will probably have to propose a new conception of this process. Schmidt (1968) said, speaking of the existence of two active longitudes: "... we also witness an ordered magnetic structure, probably a circular fluxtube, which seems to have a particularly unstable mode for two azimuthal nodes. This should have a simple explanation, but does not have it yet". We have already tried to interpret the fine structure and the relatively large variety of secondary active longitudes having different inclinations with the influence of other agents. Recently a series of papers has appeared, discussing a possible effect of Rossby-type waves on the large-scale distribution of solar magnetic fields (Starr and Gilman, 1965; Gilman, 1969a, b; Kato and Nakagawa, 1969). These waves, whose discussion would go beyond this paper, could be one of such agents.

Another agent could be, for example, the '*influence of planets*'. Although this term does not always evoke sympathetic attention, just recently, thanks to many papers (Bigg, 1967; Trellis, 1966) this question has been revived. Of course, it is still very difficult to speak about the physics of such an 'influence', though some results, for example those obtained by Ambrož (1971), have shown that there is something in it. He studied the course of the direction and the amplitude of the maximal tidal deformation force caused on the sun by six planets during the time interval 1940–1969. A comparison of his histogram of frequencies of the maxima of the curve obtained (which means the histogram of time intervals between individual maxima of the force with the power spectrum of the daily mean Zürich relative sunspot numbers (Shapiro and Ward, 1966)), or with our histogram of the rotational velocity distribution of our 'active longitudes' as they are presented by faculae, shows a striking resemblance not only in the position of the main broad maximum (27 days) but also in the positions and relative intensities of the secondary maxima. The course of the position of maxima of the investigated force in Carrington coordinates resembles very much our graphs of active longitudes, both with regard to the inclinations as well as the occurrence of two parallel strips 180° apart. A direct comparison of active longitudes with this graph gives again an interesting result: the position of maximal tidal force lies in most cases between the active longitudes.

## 4. The Sun as a Star; Solar Large-Scale Magnetic and Activity Patterns and Interplanetary Space

The regularities observed in the interplanetary magnetic fields initially evoked an interest in the large-scale patterns of solar magnetic fields (Ness and Wilcox, 1966; Wilcox, 1968). What is true of the sun and its interplanetary space is to be expected of all solar-type stars. If the sun was investigated as a star, large-scale regularities should be observed: regular variations connected with the rotation (Figure 6), short irregular variations due to large-scale patterns, and secular variations related to the cyclic and secular changes. The sun would not only appear to be a magnetic variable (Bumba *et al.*, 1967; Severny, 1969), but one would also observe variations in its

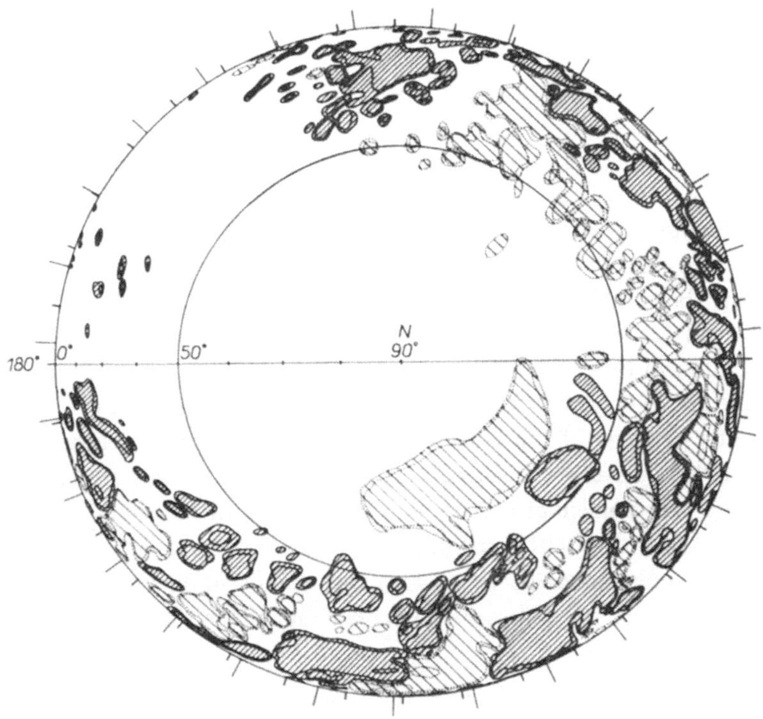


Fig. 6. The sun a magnetic variable as seen from the direction of the North pole during the rotation 1435 (December 1960, January 1961). The concentration of activity in certain longitudinal regions and the large-scale distribution of individual polarities is clearly visible.

integrated (Albrecht *et al.*, 1969) and monochromatic (Bumba and Růžičkova-Topolová, 1967) light. The close correlation of the measurements of the mean magnetic field of the Sun (seen as a star) for the period of March–June 1968 with the sector structure of the interplanetary field was recently demonstrated (Wilcox *et al.*, 1969). Such an agreement between solar and interplanetary fields could exist only if the solar source of the interplanetary field was part of a field pattern that was ordered over an appreciable portion of the solar disc. An analogy of such measurements with stellar astronomy and vice versa seems obvious, especially because the mechanism of formation of the interplanetary magnetic field sector structure is known (Schatten *et al.*, 1969; Wilcox and Colburn, 1969). The correlation of its boundaries with active longitudes was shown (Bumba and Obridko, 1969). The reflection of the large-scale pattern of solar magnetic fields in interplanetary space is such, that the rigid body rotation of the main solar activity zone crowded by magnetic fields seems also to be reflected in the corotation of sector boundaries without any dependence on heliographic latitude in a very broad interval of latitudes ($\pm 30°$) (Wilcox and Howard, 1968). Of course, the corona and the outer layers of the solar atmosphere should be influenced still more by the large-scale structures. For example a correlation between the right-handed polarization of the solar continuum radio emission and the interplanetary magnetic sectors with outward polarity was observed (Krüger *et al.*, 1968).

So far we have spoken about the longitudinal distribution of interplanetary magnetic fields. But following the close relations of both solar and interplanetary fields in the plane of the ecliptic and assuming that the magnetic fields at high latitudes are carried away by the solar wind in a way analogous to the low latitude situation, we may predict the interplanetary fields that would be observed above the ecliptic plane. We may expect that the patterns of the solar equatorial magnetic body will be reflected more or less identically up to latitudes of around 30°. Above this, the scale of the sectors may be somewhat larger and the rotational period will appear to be about 29 days. From about 40° or 50° to 65° or 70°, the fields will be quite weak and the polarity will tend to be constant with some fluctuations in amplitude over periods of months. Little can be said of still higher latitudes. Up to now space probes could not be launched with energy sufficient to reach such highly inclined orbits and make possible the investigation of the latitudinal distribution of the interplanetary fields. But comets could yield some information (Biermann and Lüst, 1963; Bappu and Sivaraman, 1969). In particular comets like Bennet's, with its orbital plane inclined by about 90° to the plane of the ecliptic may be used as natural probes for the latitudinal distribution characteristics, especially for examining the abrupt changes to be expected in latitudes around 30° in the interplanetary magnetic field.

## 5. Conclusion. Problems Connected with the Long-Term Forecasts of Solar Activity

What has been said about large-scale regularities in the solar magnetic field and activity distribution is promising for the prediction of activity on the basis of past observations when the start of a complex of activity in a stream or active longitude

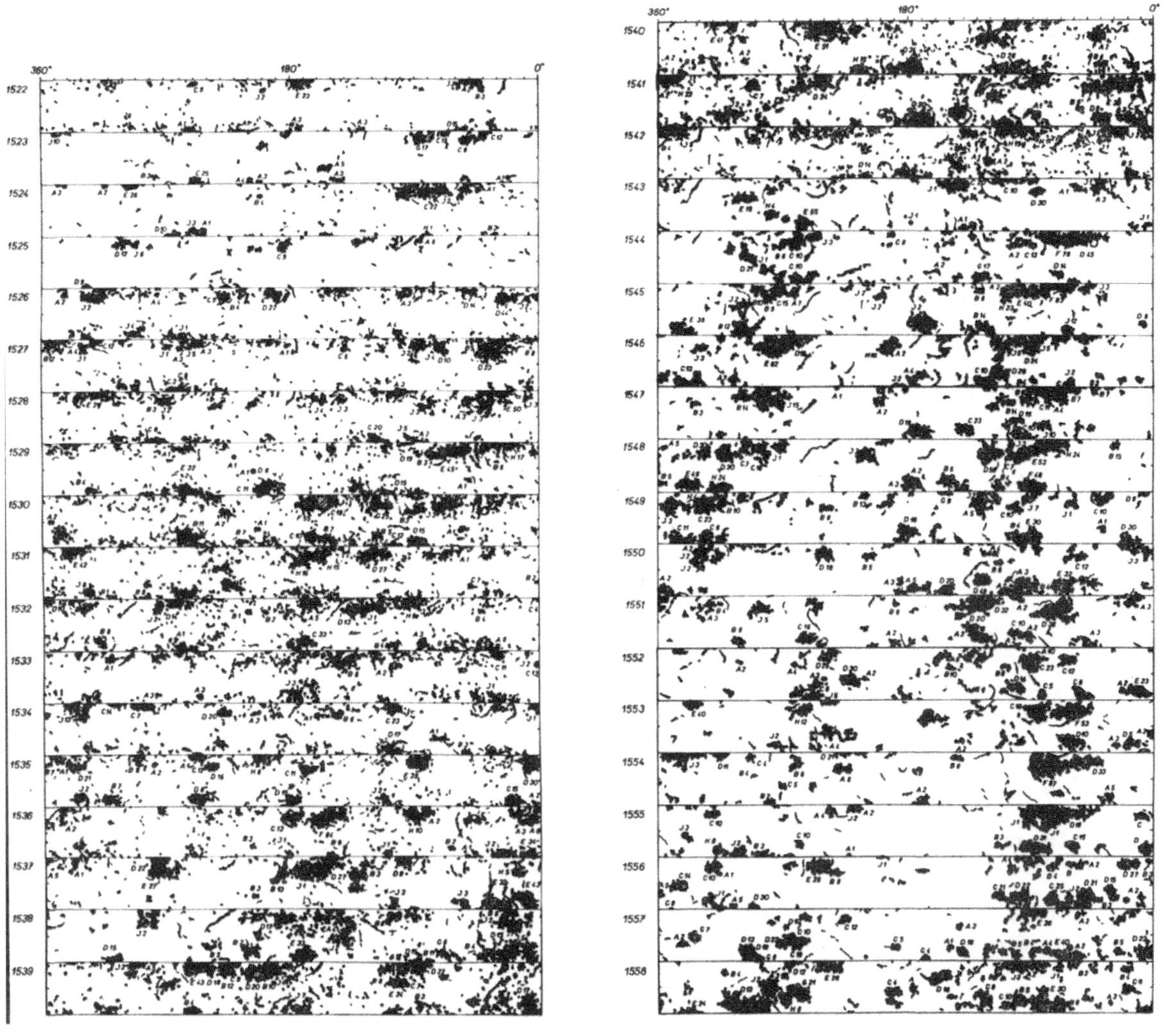

Fig. 7. Solar activity in equatorial strips (±20°) demonstrated in the daily 'Maps of the Sun' (Fraunhofer Institute) during 37 rotations (Nos. 1522–1558).

is detected. Then, the longitudinal distribution of enhanced activity over the solar disc, and the time of duration of the complex may be predicted. Of course, some forecasts concerning flare activity and geophysical consequences can also be made, but only in a very general form, unless we systematically follow the development of the activity in the complex. But all this concerns only the time period when the activity may be separated in distinct active longitudes. Furthermore, since the physical background of these regularities is not known, the fairly frequent sudden changes in the course of magnetic streams or active longitudes cannot be predicted; yet they are often accompanied by very high activity. For instance, the synoptic charts made from the daily Fraunhofer 'Maps of the Sun' cut into equatorial strips and mounted in chronological order show that for many rotations through the years 1968 and 1969 the concentration of activity in active longitudes was very regular (Figure 7). But in the period August–September 1969 (rotation 1551-2) the situation began to change and such a change would probably be very difficult to predict.

Obviously, many investigations are needed. Above all, the theoretical interpretation of the observed morphological regularities is insufficient. Because of the practical significance of predictions of 'cosmic weather', more scientists might soon be interested in the problem.

## References


Albrecht, R., Maitzen, H. M., and Rakos, K. D.: 1969, *Astron. Astrophys.* **3**, 236.
Ambrož, P.: 1971, in press.
Antalová, H. and Gnevyshev, M. N.: 1965, *Astron. Zh.* **42**, 253.
Balusubrahmanyan, V. K. and Venkatesan, D.: 1970, *Solar Phys.* **11**, 151.
Bappu, M. K. V. and Sivaraman, K. R.: 1969, *Solar Phys.* **10**, 496.
Bigg, E. K.: 1967, *Astron. Zh.* **72**, 463.
Biermann, L. and Lüst, R.: 1963, in *The Moon, Meteorites, and Comets* (ed. by Middlehurst and Kuiper), Chicago, p. 618.
Bumba, V.: 1967, *Rendiconti della Scuola Internazionale di Fisica "E. Fermi"*, XXXIX Corso, 77.
Bumba, V.: 1970, *Solar Phys.* **14**, 80.
Bumba, V. and Howard, R.: 1965a, *Astrophys. J.* **141**, 1502.
Bumba, V. and Howard, R.: 1965b, *Astrophys. J.* **141**, 1492.
Bumba, V. and Howard, R.: 1969a, in C. de Jager and Z. Švestka (eds.), 'Solar Flares and Space Research', *XIth Cospar Symposium*, North-Holland Publ. Co., Amsterdam.
Bumba, V. and Howard, R.: 1969b, *Solar Phys.* **7**, 28.
Bumba, V. and Letfus, V.: 1971, in press.
Bumba, V. and Obridko, V. N.: 1969, *Solar Phys.* **6**, 104.
Bumba, V. and Růžičková-Topolová, B.: 1967, *Solar Phys.* **1**, 216.
Bumba, V. and Růžičková-Topolová, B.: 1969, *Bull. Astron. Inst. Czech.* **20**, 63.
Bumba, V., Howard, R., and Smith, S. F.: 1964, *Carnegie Inst. of Washington Year Book* **63**, 6.
Bumba, V., Howard, R., and Smith, S. F.: 1967, in R. C. Cameron (ed.), *The Magnetic and Related Stars*, Mono Book Corp., Baltimore, p. 131.
Bumba, V., Howard, R., Martres, M. J., and Soru-Iscovici, I.: 1968, in K. O. Kiepenheuer (ed.), 'Structure and Development of Solar Active Regions', *IAU Symp.* **35**, 13.
Bumba, V., Howard, R., Kopecký, M., and Kuklin, G. V.: 1969, *Bull. Astron. Inst. Czech.* **20**, 18.
Dodson-Prince, H. W. and Hedeman, E. R.: 1968, in K. O. Kiepenheuer (ed.), 'Structure and Development of Solar Active Regions', *IAU Symp.* **35**, 56.
Gilman, P. A.: 1969a, *Solar Phys.* **8**, 316.
Gilman, P. A.: 1969b, *Solar Phys.* **9**, 3.
Gnevyshev, M. N.: 1963, *Astron. Zh.* **40**, 401.
Gnevyshev, M. N.: 1967, *Solar Phys.* **1**, 107.
Howard, R., Bumba, V. and Smith, S. F.: 1967, Carnegie Inst. of Washington, Publ. No. 626, Washington.
Kato, S. and Nakagawa, Y.: 1969, *Solar Phys.* **10**, 476.
Krüger, A., Bumba, V., Howard, R., and Kleczek, J.: 1968, *Bull. Astron. Inst. Czech.* **19**, 180.
Levitskij, L. S.: 1967, *Izv. Krymsk. Astron. Observ.* **37**, 137.
Livingston, W. C.: 1969, *Solar Phys.* **7**, 144.
Ness, N. F. and Wilcox, J. M.: 1966, *Astrophys. J.* **143**, 23.
Schatten, K. H., Wilcox, J. M., and Ness, N. F.: 1969, *Solar Phys.* **6**, 442.
Schmidt, H. U.: 1968, in K. O. Kiepenheuer (ed.), 'Structure and Development of Solar Active Regions', *IAU Symp.* **35**, 71.
Shapiro, R. and Ward, F.: 1966, *J. Geophys. Res.* **71**, 2385.
Severny, A.: 1969, *Nature* **224**, 53.
Simon, G. W. and Weiss, N. O.: 1968a, *Z. Astrophys.* **69**, 435.

Simon, G. W. and Weiss, N. O.: 1968b, in K. O. Kiepenheuer (ed.), 'Structure and Development of Solar Active Regions', *IAU Symp.* **35**, 108.
Starr, V. P. and Gilman, P. A.: 1965, *Tellus* **17**, 334.
Sýkora, J.: 1969, *Bull. Astron. Inst. Czech.* **20**, 70.
Švestka, Z.: 1966, *Bull. Astron. Inst. Czech.* **17**, 262.
Švestka, Z.: 1968a, *Solar Phys.* **4**, 18.
Švestka, Z.: 1968b, in K. O. Kiepenheuer (ed.), 'Structure and Development of Solar Active Regions', *IAU Symp.* **35**, 287.
Švestka, Z. and Simon, P.: 1969, *Solar Phys.* **10**, 3.
Trellis, M.: 1966, *Compt.-Rend.* **262**, 312, 376.
Vitinskij, Ju. I.: 1969a, *Solar Phys.* **7**, 210.
Vitinskij, Ju. I.: 1969b, *Sol. Dann.*, No. 4, 88.
Warwick, C. S.: 1965, *Astrophys. J.* **141**, 500.
Wilcox, J. M.: 1968, *Space Sci. Rev.* **8**, 258.
Wilcox, J. M. and Colburn, D. S.: 1969, *J. Geophys. Res.* **74**, 2388.
Wilcox, J. M. and Howard, R.: 1968, *Solar Phys.* **5**, 564.
Wilcox, J. M., Severny, A., and Colburn, D. S.: 1969, *Nature* **224**, 353.
Wolfer, A.: 1897, *Publ. Sternw. Eidg. Polytechnikums Zürich* **1**, XII.

# LOCAL MAGNETIC FIELDS ON THE SUN

A. B. SEVERNY

*Crimean Astrophysical Observatory, U.S.S.R.*

The strongest magnetic fields on the sun are those of sunspots. Although their magnetic fields have been studied for a long time much remains unknown about the magnetism of these principal sources. At sufficiently large distances from a spot the magnetic field can, apparently, be well approximated as that of a dipole submerged below the solar photosphere, (classical Mt. Wilson observations of Hale, 1919; Bumba, 1960; Severny, 1964). However, a more detailed study of the distribution of magnetic fields inside the spot shows the fine structure of the field. The field is concentrated in bundles, small elements, showing twisting and vorticities, somewhere inside a sunspot (Severny, 1964; Stepanov and Gopasyuk, 1962). The field is also appreciably inhomogeneous, and has fine structure shown by a local maximum of the field strength at a certain depth (Guseeynov, 1969), the strong rotation of the transverse component of the magnetic field with depth (Severny, 1964), the fine structure in the pattern of vertical electric currents (Severny, 1964; Kotov, 1970).

These findings are in accordance with the recently found concentration of strong magnetic fields in small portions of the solar plasma. Steshenko (1967) found that even in small pores of 2″ the magnetic field can reach values of 1000 gauss and more. Beckers and Schröter (1968) and Sheeley (1967) found structures like "microspots" – fine structural elements of the magnetic field which can be observed in large numbers only by means of spectra of active regions surrounding the visible sunspots. It seems not impossible that the frequently observed bias of the flux of one polarity in sunspot groups is compensated by an opposite (integral) flux from these elements. It is also not excluded that sunspots are complexes of such microspots.

Although an azimuthal component of the magnetic field is sometimes observed around sunspots (Stepanov and Gopasyuk (1962)), Severny (1964), Kotov (1970)) the observed distribution and structure of the field of sunspots can hardly be similar to a force-free field, because for the real field the condition

$$\mathbf{j} \times \mathbf{H} = 0, \quad \mathbf{j} = \frac{c}{4\pi} \operatorname{curl} \mathbf{H}$$

is not fulfilled (nevertheless the approximation $\mathbf{j} = 0$ is quite acceptable away from sunspots). The Hα-vortex structure around sunspots (Hale vortices) is of peculiar interest, because there is a close correspondence between this structure and the behaviour of the transverse component of the magnetic fields; this relates also to the vortex structure observed in white light in penumbral regions (Tsap, 1963; Stepanov, 1958; Steshenko, 1969). Kotov (1970) found that such vortex structures of the transverse field can at the same time also be observed in the deeper layers of a sunspot. The sense of spiraling in Hα is usually the same in the northern hemisphere and

*Dyer (ed.), Solar Terrestrial Physics / 1970: Part I, 38–48.* 

opposite to that in the southern hemisphere of the sun; it depends neither on the magnetic polarity, nor on the cycle of solar activity. The explanation of these structures should therefore be similar to that of terrestrial cyclones. In that case, the origin of the vortex structures of the magnetic fields in sunspots is of a pure hydrodynamical nature, and the magnetic field follows the motion of the gases and cannot be the primary cause of these vortices. The magnetic field can only influence the transport of heat by inhibition of the strong convection, so providing a reasonable explanation for the darkness of sunspots. The field strength needed for such an inhibition is probably slightly larger than 1000 gauss, because the small elements with fields $\leqslant 10^3$ gauss are frequently observed without any darkening (Steshenko, 1967).

Attempts to explain the formation of a sunspot as the result of a redistribution or an assembling of magnetic fields from a large area by convection or supergranulation in the lower layers of the convective zone are hardly consistent (Schmidt, 1968; Weiss, 1964; De Jager, 1964). It is difficult to imagine how convection can produce the observed predominance of bipolar groups in a definite range of latitudes on the sun (activity zones), and secure a definite sequence of polarities such as e.g. a N-polarity in leading, and a S one in following sunspots (or the reverse), and to produce also a change of polarities with the 11-year cycle. "We cannot get order out of disorder" as was stated quite reasonably by Cowling (1966). Supergranulation and convection of different characteristic dimensions cannot play any creative role, but plays rather a destructive part, because the main process connected with it is to break into pieces and to twist the tubes of field lines of different polarities, thus leading to the above-mentioned fine structural field-pattern superimposed on the quasi-dipolar main background field, (see Pikelner 1966). This follows also from a consideration of the motions superimposed on the regular motions in sunspot groups – viz. the effect of two-dimensional diffusion with the coefficient $D = \frac{1}{2}lV = 10^9 \times 10^4 = 10^{13}$ $cm^2/sec$.

The tubes of field lines like long snakes or like toroidal structures must pre-exist in the lower parts of the solar atmosphere independently of convection. This is supported by the increase of the transverse field with depth, as observed in unipolar sunspots (Kotov, 1970). A reasonable model of a unipolar sunspot would be an electric ring current, or a current forming a spiral with a steep increase outwards, thus producing a solenoidal field downwards and a quasi force-free field in the upper layers ($\mathbf{j}$ being parallel to the direction of the magnetic field). Recent measurements by Kotov of the transverse fields at different levels of unipolar spots favour such a suggestion (see Figure 1).*

Descending motions are observed above sunspots (Gopasyuk, 1967), and longitudinal motions are frequently observed in those parts of active regions where the field is mostly transverse. Conversely, transverse motions are characteristic for regions of longitudinal fields (Gopasyuk, 1968). Motions across the magnetic field are also

* The measurements made in SIBIZMIR showing that the transverse field appears first during the development of a group are also in agreement with this idea (Bappu *et al.*, 1968).

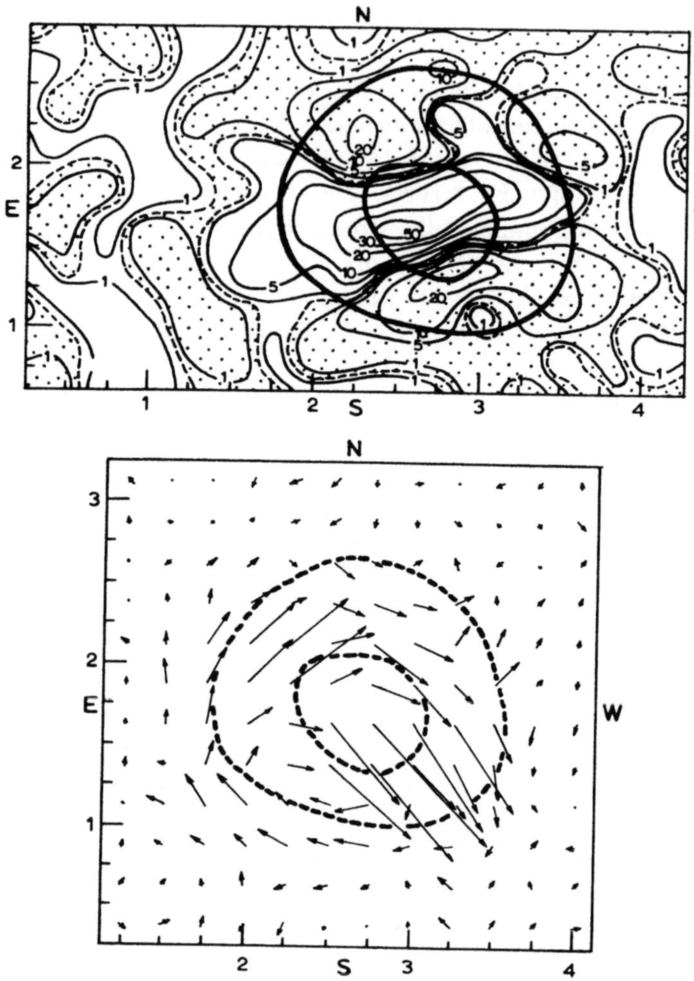


Fig. 1. Distribution of vertical (upper diagram) and horizontal (lower diagram) electrical currents in an unipolar sunspot. The scale unit of the two diagrams is 25″. The unit of electrical current density is $10^{-2}$ gauss/km; see Kotov (1970).

observed in investigations of the Evershed effect in Cambridge (Holmes, 1961) and in Crimea (Gopasyuk, 1968).

Passing to the magnetic field of the whole active region, or to characteristic dimensions $\approx 10^9$–$10^{10}$ cm we should first pay attention to the supergranulation (Leighton, 1963), which is considered nowadays an important factor in many solar phenomena. Although the transverse velocity elements show a well-defined structure with characteristic dimensions around 30–40 thousand km, well defined magnetic structures of the same dimensions can not be found (see Figure 2). The histogram shows only a slight increase of the number of magnetic elements near the characteristic size of supergranules (45″). Moreover, a recent very careful statistical comparison of maps of the line-of-sight velocities $V_{\parallel}$ with those of longitudinal fields $H_{\parallel}$ showed that the maximal fields as well as the corresponding brightest regions of the chromosphere are located immediately near or on the line where the longitudinal velocity $V_{\parallel}=0$; see Figure 3 (Gopasyuk and Tsap, 1969). This holds for the center as well as for the limb region of the sun as well as for active or quiet regions. Meanwhile, according to the concept of supergranulation the maximal fields must lie in areas of maximal

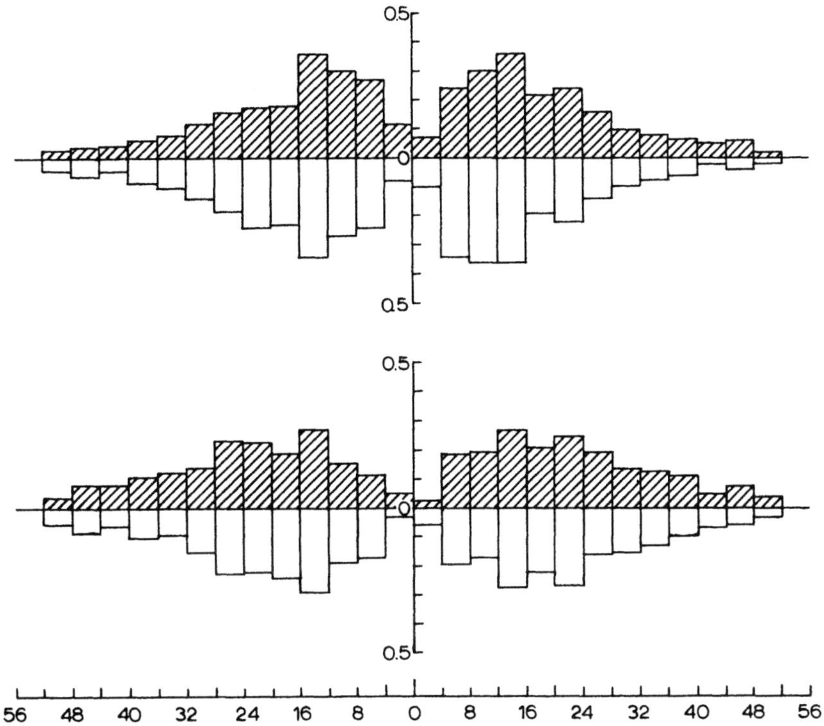


Fig. 2. Histograms (percentages) of the sizes of magnetic features outside active regions in polar (upper diagram) and equatorial regions (lower diagram). Shaded histograms refer to the north polar region and to areas north of the equator, the others refer to the southern regions. North polarity elements are drawn on the left side, south polarities are to the right. The horizontal scale is given in seconds of arc.

descending velocities. At the same time Bumba and Howard (1965) and Bappu *et al.* (1968) found, that not only flocculi but also sunspots appear in the corners or at the borderlines of adjacent supergranulation cells. These results probably reflect the fundamental property of magnetic structures on the sun, namely the close correlation between magnetic fields, and in particular between their strength, and the Hα- and calcium K-line-chromospheric structure, showing that the emission intensity in these lines increases statistically with the (longitudinal or transverse) field strength components. Tsap (1969) showed that this dependence holds also for photospheric metallic lines. It is also not excluded that the magnetic fields, which appear before the emission, and disappear after all other features of an active center, can also be connected with the regions of enhanced radioemission at short wavelengths (cm and mm range) as well as with regions of enhanced X-ray emission on the sun. This can be inferred from a comparison of magnetic maps of the whole disk with the distribution of radioemission at 1, 2, 8 mm and 3 cm made in Crimea (Efanov *et al.*, 1968).

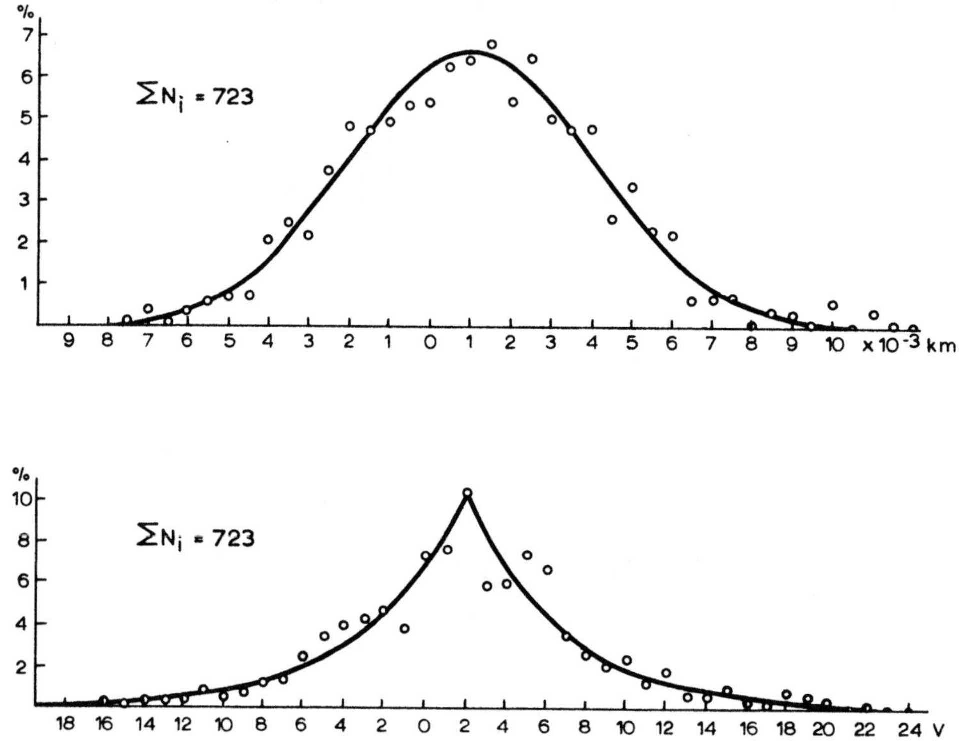


Fig. 3. Distribution of the shortest distances (in thousands of km) between the line $V_{\|} = 0$ and the points of maximal field strength $H_{\|}$ in areas of ascending motions (left side) and of descending motions (right). One unit along the abscissa corresponds to 29 m/sec.

The close correlation between the emission and magnetic fields has been the subject of many theoretical considerations. Pikelner (1960) pointed out that increasing magnetic field strength can suppress turbulence more easily than convection, and this can make heat transfer more efficient in regions of enhanced fields. But this concept now meets difficulties, first as mentioned above, since flocculi (as well as flares (Severny, 1960)) tend to appear in regions where the longitudinal motions are slow or entirely absent. The second difficulty is that the brightness of flocculi (in H$\alpha$ and K-light) increases not only with $H_{\|}$, but also with the transverse field strength $H_{\perp}$ (there is no maximum in the dependence of brightness on field strength).

In regions of enhanced magnetic fields magneto-acoustic waves can appear and the strong damping of these waves can be a source of heat (Piddington, 1956). At the same time high-frequency plasma oscillations can also appear in these regions, and these oscillations can, in principle, also heat the plasma; the enhancement of short-wave radioemission in these regions points to this possibility. In other words, the heating can come from above if magnetic fields are extending sufficiently far upward. The connection of magnetic fields in the lower and upper parts of the solar atmosphere is probably very close, as can be seen from a comparison of magnetic maps recorded in $\lambda 5250$ and H$\alpha$, for instance. The similarity is striking; at least 90% of the features are the same. In undisturbed regions with weak fields, the magnetic fields are carried into the corona and interplanetary space by convective motions and the solar wind. The very satisfactory agreement between the magnitude and the sign of the interplanetary fields and the general field of the sun is evidence that photospheric fields are carried not only into the upper corona, but also into the interplanetary space from regions outside active regions (mechanism developed by Schatzman (1962) and Mestel (1968)).

The correspondence between chromospheric structures and magnetic fields can even be traced outside active regions (Stepanov, 1958; Zirin and Werner, 1967). Formations like filaments and luminous threads connect regions of opposite polarities and can be as long as 100000 km and more. Nevertheless, the relation between fields and chromospheric structures is a rather qualitative one. From the fact that these structures sometimes do not change greatly, no conclusion can be drawn about the behaviour of magnetic fields as was done by Zirin and Werner (1967): at least a careful photometric analysis of many chromospheric formations is necessary.

The overall magnetic field structure between sunspots in an active region does not seem very different from the structure resulting from a system of dipoles, equivalent to the system of the spots (Gopasyuk *et al.*, 1963; Severny, 1963; Godovnikov and Smirnova, 1965; Künzel, 1969), except for the cases of the so-called 'wrong connections', when the lines of force are not closed at the nearest 'pole', but at another, more distant 'pole', and an apparent crossing of the lines of force occurs, due to the interaction of several opposite polarities. We should keep in mind that in the case of solar active regions we are in the very unfavourable intermediate position, that the magnetic energy $H^2/8\pi$ is comparable with the kinetic energy of motions and with the heat energy, and this makes theoretical consideration here especially difficult.

Not too much is known about the early development of active regions. The observations made in SIBIZMIR (Bappu *et al.*, 1968) showed that the appearance of a transverse field precedes the sunspots and is probably characteristic for the early stage of the development. It looks as if the edge of a toroidal bundle or a ring of field lines emerged. A bias toward one polarity is characteristic for the early stage of the development of active regions according to Bumba and Howard (1965). The growth of longitudinal fields begins in the center of a sunspot, and is accompanied by the inflow of gases into the developing spot (Gopasyuk, 1967). The decay of the magnetic regions is accompanied by a breaking of these regions into smaller ones, by a decrease of the magnetic flux and by another violation of the balance of the fluxes (Gopashuk, 1967). However, sometimes a new region of opposite polarity can emerge inside a decaying magnetic region; this process is probably characteristic for active regions and is accompanied by a rise of activity (Bumba and Howard, 1965;

the observations of the Meudon astronomers: Martres *et al.*, 1968). A similar process is also observed for weak fields when recorded with a high time-resolution (see Figure 4).

Let us consider, in conclusion, the role of magnetic fields in non-stationary processes and especially in solar flares. At present there is no doubt about the close connection between solar, esp. flare, activity with magnetic fields. The appearance of flares in specific parts of magnetic regions, such as regions of highest field gradient, or regions near the neutral line $H_{\|}=0$, or on the apparent crossing of the transverse fields $H_{\perp}$ or points of contact of oppositely directed field vectors $H_{\perp}$, is suggestive. Also very characteristic is the connection of the first brightening of flares with maximal

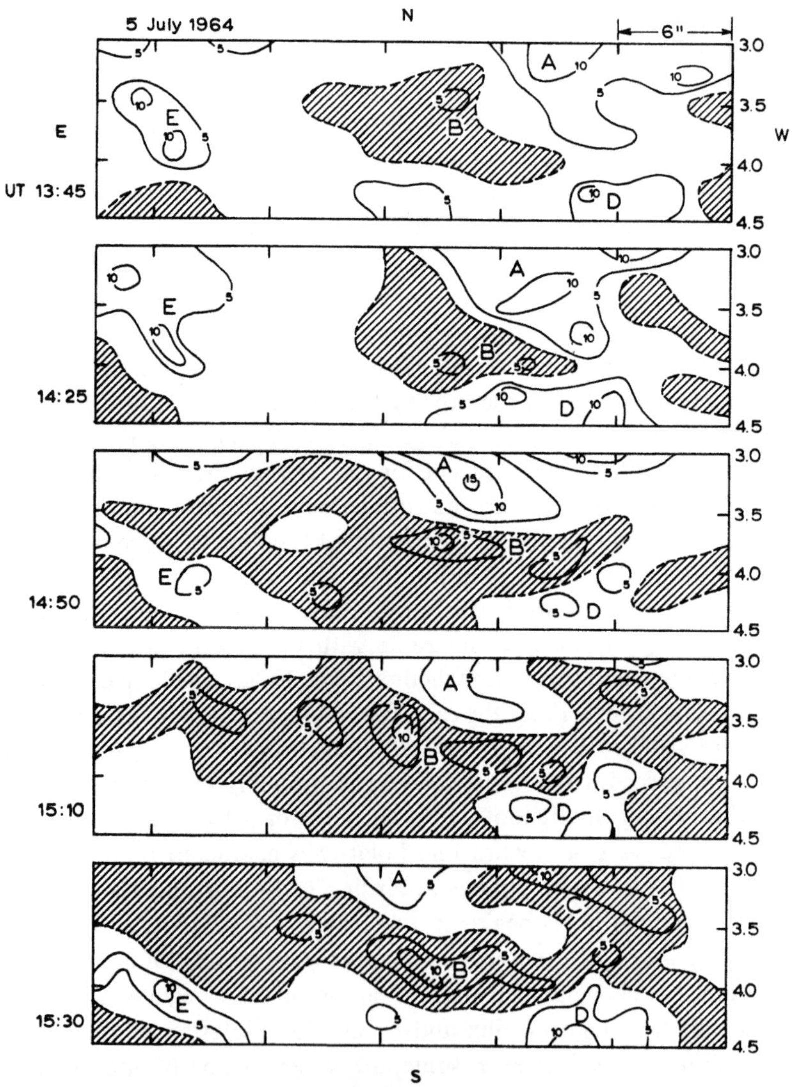


Fig. 4. Illustrating the process of emerging of one polarity (shaded) into a region of opposite polarity, in the course of time. The lines are labeled with their field strength in gauss.

vertical electric currents as well as with the junction of oppositely directed vertical or horizontal electric currents. This last feature is the other side of the well known property of flares to appear in places where the gradients of the field exceed a 'critical' value (Severny, 1968; Smith and Ramsey, 1967; Rust, 1968; and others). It is also difficult not to accept the conclusion about appreciable, at least temporary, changes of magnetic fields connected with solar flares. Figure 5 illustrates the changes of the gradients and the field energy of the whole active region and the changes of the net magnetic flux through the active region – all connected with the well-known proton flare of 17 July 1966. These changes are apparently more clearly visible for large

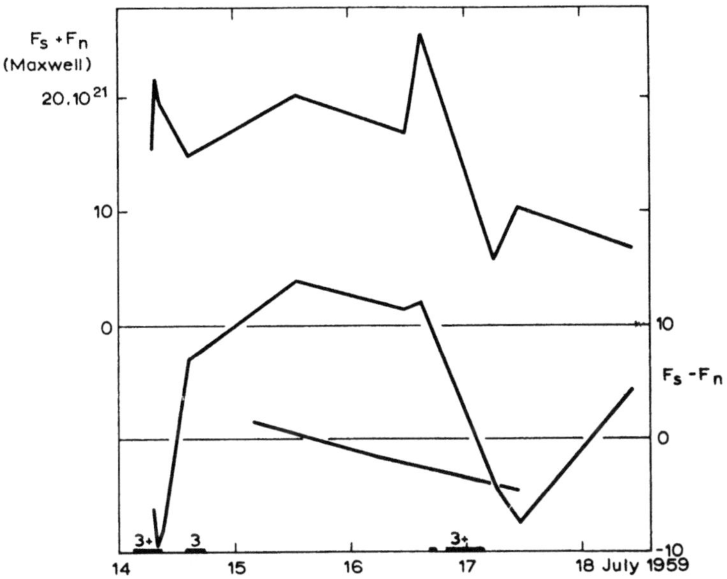


Fig. 5. Upper diagram: Variation with time of the total flux ($F_S + F_N$, in maxwells) associated with the great 3+ flare of July 17, 1959. Lower diagram: The same for the net flux ($F_S - F_N$). The lower line is the flux for sunspots alone, according to routine observations of maximal magnetic fields in umbrae, multiplied with the spot area.

flares; in the smaller ones the effect is not well visible or absent. The observation that there are no changes of chromospheric structure (from visual inspection) in the H and K-line is considered by some authors (Zirin and Werner, 1967; Bappu *et al.*, 1962) as evidence against the change of magnetic fields connected with flares. This conclusion is, however, incorrect because it is based on qualitative inspections of the chromospheric pattern (which, by the way, sometimes shows considerable changes). In fact, the chromospheric pattern in active regions remains frequently without essential changes, as was for instance the case for the flare-active group of 13–16 September 1963 (Morton and Severny, 1968) and during the great proton flare of 17 July 1959. For the last flare not only the chromospheric pattern but also the qualitative magnetic pattern remained unchanged (see Babcock and Howard, 1960).

However a more careful *quantitative* analysis of the photospheric magnetic maps of these groups showed very pronounced changes of the energy and of the fluxes connected with flares (Howard and Severny, 1963).

We also wish to mention the recently observed relation between flares and the change of the net magnetic flux through the active region, a phenomenon observed in Meudon (Martres *et al.*, 1968) and Crimea (Severny, 1969). This change of the flux could be connected with the appearance of an electromotive force through the boundary of the region, provided that there are no motions which may violate this simple consideration (and produce an electromotive induction force $(1/c)\,\mathbf{v}\times\mathbf{H}$). Estimates show that these motions are small and the corresponding electro motive force is negligible (not larger than 10%) as compared with the value in a static contour. It is remarkable that this force was $10^8$–$10^9$ eV for the proton flares of 7 July and 2 Sept., 1966, hence was comparable with the energy of cosmic rays generated by proton flares. The important question is the amount of the contribution of the electromotive induction force. Strictly, the change of the magnetic flux must be calculated through the moving boundary, rigidly connected with the particles, and this change is zero in the case of infinite conductivity. Meanwhile the imbalance of fluxes is quite a common phenomenon during the development of an active region. In order to obtain agreement between the electric currents

$$\mathbf{j} = \frac{c}{4\pi}\,\mathrm{curl}\,\mathbf{H} = \sigma\left(\mathbf{E} + \frac{1}{c}\,\mathbf{v}\times\mathbf{H}\right)$$

found from the measurements of **H** and curl **H** with the value

$$c\ \varphi\,(\mathbf{E}\,\mathrm{d}\mathbf{s}) = -\frac{\partial}{\partial t}\int\!\!\int H_n\,d\sigma$$

found also from the measurements of the net magnetic flux (from maps of the longitudinal fields $H_{\parallel}$) we must reduce the conductivity $\sigma$ by a factor $10^4$. Hence the value of $\mathrm{d}H_n/\mathrm{d}t$ through the moving boundary does not vanish. For large flares the induction term is probably negligible, and one may consider the flare phenomenon as connected with strong electric currents according to the theory of flares developed by Alfvén and Carlquist (1967).

There is little correspondence between the positions of flares with the neutral points of the total magnetic vector, and this creates difficulties for the neutral-point theory of flares. It is possible, however, to avoid this difficulty by the assumption that the correspondence exists only for the higher layers which can not be observed. Some information can be obtained on this point from simultaneous measurements of fields at the photospheric ($\lambda$5250) and chromospheric (H$\alpha$) levels. Figure 6 shows such simultaneous recordings and they suggest the possibility of a configuration of the field in which a loop of the field emerges from the photosphere with opposite direction of the transverse field. Such a field situation (bifurcations) can also be observed in the horizontal plane (e.g. in the case of the flare of 7 July, 1966). Such loops can represent the upper part of a toroid pushed up from below by buoyancy forces. Such fields can probably exist separately in the chromosphere (not having their roots in the photosphere), as some observations show (Mogilevsky and Shelting, 1964; Severny, 1966). The possibility that such detached chromospheric magnetic configurations could exist is also suggested by field observations in prominences (Zirin, 1961; Ioshpa 1962; Kotov, 1969).

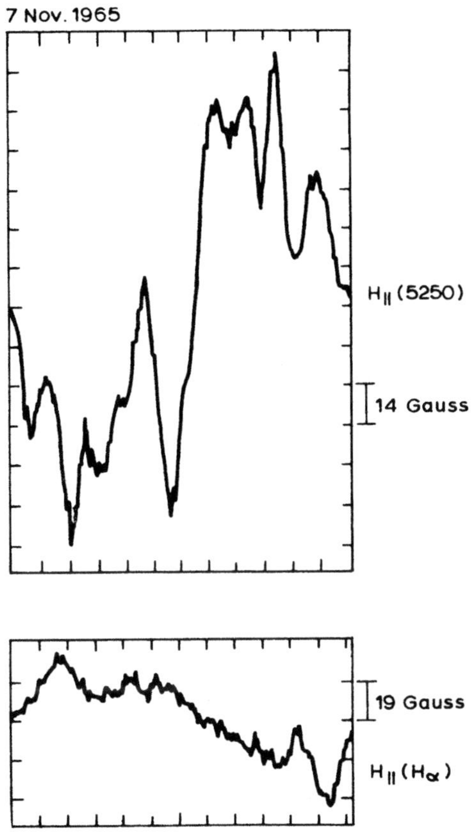


Fig. 6.

Independently, however, from a theory of flares, the appearance of flares in specific field configurations (high gradients and electric currents) permits us to predict flares using magnetic field data. The high percentage (75–80%) of correct forecasts shows also the practical use of field data for manned space flight as well as for supersonic aviation.

## References

Alfvén, H. and Carlquist, P.: 1967, *Solar Phys.* **1**, 220.
Bappu, M. K. V., Bhatnagar, A. and Punetha, L.: 1962, *Observatory* **82**, 170.
Bappu, M., Grigorjev, V., and Stepanov, V.: 1968, *Solar Phys.* **4**, 409.
Beckers, J. and Schröter, E.: 1968, *Solar Phys.* **4**, 142.
Bumba, V.: 1960, *Izv. Krymsk. Astrofiz. Observ.* **23**, 212.

Bumba, V. and Howard, R.: 1965, *Astrophys. J.* **141**, 1492.
Cowling, T.: 1966, in R. Lüst (ed.), 'Stellar and Solar Magnetic Fields', *IAU Symp.* **22**, 405.
De Jager, C.: 1964, *Bull. Astron. Inst. Neth.* **17**, 253.
Efanov, V., Moiseev, I., and Severny, A.: 1968, in K. O. Kiepenheuer (ed.), 'Structure and Development of Solar Active Regions', *IAU Symp.* **35**, 588.
Efanov, V., Kislyakov, A., Moiseev, I., and Naumov, I.: 1969, *Solar Phys.* **8**, 331.
Godovnikov, N. and Smirnova, E.: 1965, *Izv. Krymsk. Astrofiz. Observ.* **33**, 86.
Gopasyuk, S.: 1967, *Izv. Krymsk. Astrofiz. Observ.* **36**, 56.
Gopasyuk, S.: 1967, *Izv. Krymsk. Astrofiz. Observ.* **37**, 29.
Gopasyuk, S.: 1968, *Izv. Krymsk. Astrofiz. Observ.* **38**, 65.
Gopasyuk, S. and Tsap, T.: 1969, *Astron. Zh.* **46**, 923; 1969, *Izv. Krymsk. Astrofiz. Observ.* **44**.
Gopasyuk, S., Ogir, M., Severny, A., and Schaposschnikova, E.: 1963, *Izv. Krymsk. Astrofiz. Observ.* **29**, 15; see also Smirnova, N. and Godovnikov, N.: 1964, *Krymsk. Astrofiz. Observ.* **33**, 86.
Gusseynov, M.: 1969, *Izv. Krymsk. Astrofiz. Observ.*, **41–42**, 253.
Hale, G. E. *et al.*: 1919, *Astrophys. J.* **59**, 153.
Holmes, J.: 1961, *Monthly Notices Roy. Astron. Soc.* **122**, 301.
Howard, R. and Babcock, H. W.: 1960, *Astrophys. J.* **132**, 218.
Howard, R. and Severny, A.: 1963, *Astrophys. J.* **137**, 1241.
Joshpa, B.: 1962, *Geomagn. Aeronomija* **2**, 172.
Kotov, V.: 1969, *Izv. Krymsk. Astrofiz. Observ.* **39**, 276.
Kotov, V.: 1970, *Izv. Krymsk. Astrofiz. Observ.* **41–42**, 67.
Künzel, H.: 1969, *Astron. Nachr.* **291**, 265.
Leighton, R.: 1963, *Ann. Rev. Astron. Astrophys.* **1**, 19.
Martres, M., Michard, R., Soru-Iscovici, I., and Tsap, T.: 1968, *Solar Phys.* **5**, 187.
Mestel, L.: 1968, *Monthly Notices Roy. Astron. Soc.* **138**, 359.
Mogilevsky, E. and Shelting, B.: 1966, in M. Cimino (ed.), *Proceedings of the Meeting on Solar Magnetic Fields, and High Resolution Spectroscopy*, Roma, p. 222.
Moreton, G. and Severny, A.: 1968, *Solar Phys.* **3**, 282.
Piddington, J.: 1956, *Monthly Notices Roy. Astron. Soc.* **116**, 314.
Pikelner, S.: 1960, *Astron. Zh.* **37**, 616.
Pikelner, S.: 1966, *Usp. Fiz. Nauk.* **88**, N3, 505.
Rust, D. M.: 1968, in K. O. Kiepenheuer (ed.), Structure and Development of Solar Active Regions', *IAU Symp.* **35**, 77.
Schatzman, E.: 1962, *Ann. Astrophys.* **25**, 1.
Schmidt, H. U.: 1968, in K. O. Kiepenheuer, (ed.), 'Structure and Development of Solar Active Regions', *IAU Symp.* **35**, 95.
Severny, A.: 1960, *Izv. Krymsk. Astrofiz. Observ.* **24**, 281.
Severny, A.: 1963, *Izv. Krymsk. Astrofiz. Observ.* **30**, 161.
Severny, A.: 1964, *Izv. Krymsk. Astrofiz. Observ.* **33**, 34.
Severny, A.: 1966, *Astron. Zh.* **43**, 465.
Severny, A.: 1969, in C. de Jager and Z. Švestka (eds.), 'Solar Flares and Space Research', *Cospar Symp.*, p. 38.
Severny, A.: 1969, *Ann. IQSY* **3**, 11.
Severny, A. and Zvereva, A.: 1969, *Izv. Krymsk. Astrofiz. Observ.* **44**.
Sheeley, N.: 1967, *Solar Phys.* **1**, 171.
Smith, S. and Ramsey, H.: 1967, *Solar Phys.* **2**, 158.
Stepanov, V.: 1958, *Izv. Krymsk. Astrofiz. Observ.* **20**, 52.
Stepanov, V. and Gopasyuk, S.: 1962, *Izv. Krymsk. Astrofiz. Observ.* **28**, 194
Steshenko, N.: 1967, *Izv. Krymsk. Astrofiz. Observ.* **37**, 21.
Steshenko, N.: 1969, *Izv. Krymsk. Astrofiz. Observ.* **39**, 245.
Tsap, T.: 1963, *Izv. Krymsk. Astrofiz. Observ.* **31**, 200.
Tsap, T.: 1969, *Izv. Krymsk. Astrofiz. Observ.* **39**, 265.
Weiss, N.: 1964, *Monthly Notices Roy. Astron. Soc.* **128**, 225.
Zirin, H.: 1961, *Astron. Zh.* **38**, 861.
Zirin, H. and Werner, S.: 1967, *Solar Phys.* **1**, 66.

## PROPERTIES OF SOLAR ACTIVE REGIONS*

A. BRUZEK
*Fraunhofer Institute, Freiburg i.B., Germany*

### 1. Introduction

'Solar activity' is a rather general term for quite a number of different phenomena occurring in the photosphere, in the chromosphere and in the corona. These 'active' phenomena are: (1) structures such as spots, plages, filaments, flares and coronal condensations; (2) gradual and impulsive enhancements of radiation in the EUV, X-ray and radio region of the solar spectrum; (3) motions and ejections of solar plasma and the acceleration of fast electrons and protons. All these phenomena are closely interrelated and concentrated in limited regions on the Sun which usually are referred to as 'active regions'.

The fundamental physical condition of an active region is an enhanced magnetic field which – in various ways – controls or even produces the observed phenomena. In the photosphere concentrations of magnetic flux with field strengths $>1200$ G give rise to the dark sunspots; extended magnetic fields of some tens up to several hundred gauss which permeate the chromosphere and the corona are responsible for the plage phenomena, the coronal condensation, and the associated enhanced radiations; magnetic fields support and shape prominences and play a determining role in flare occurrence. Therefore 'magnetic region' is another, more pertinent and more technical term for active region.

In the present review a few aspects only of solar active regions can be discussed. These are as follows: (1) appearance of new active regions, (2) strong field phenomena (spots etc.), (3) moderate field phenomena (plages, coronal condensations and their emissions), and (4) flares.

### 2. General Conditions of Appearance and First Development of New Active Regions

There are observations indicating that the appearance and development of new active regions is not an isolated and independent process but depends on the magnetic history of the background. Bumba and Howard (1965) and Bappu *et al.* (1968) observed that new active regions tend to appear in patterns of old magnetic fields which are the remnants of previous magnetic regions. It is remarkable that the new magnetic region does not disturb the distribution of the old field. That means that the new region is not formed by a reorganization or a concentration of the old surface fields but rather that the new fields emerge from a subsurface source (Bumba, 1967; Bumba *et al.*, 1968).

A strong correlation between active region appearance and development and already

*Mitteilung aus dem Fraunhofer Institut No. 101.

*Dyer (ed.), Solar Terrestrial Physics / 1970: Part I, 49–60.* 

existing *enhanced* magnetic regions (plages, spot groups) has been established by a number of authors. New observations of Bumba and Howard (1965) and Dodson and Hedeman (1968) prove the existence of preferred zones of activity which had been inferred before by Losh (1939), Becker (1955), Martres and Michard (1965), Haurwitz (1968) and others. They found that active regions tend to form in long lived *families* or '*complexes*' of activity at preferred solar longitudes. New regions appear on the borders of older regions so that the 'complex' appears to expand; the rotation period of the W border is 27 days, that of the E border is 28–29 days (Bumba, 1972). Martres (1968, 1970) has found that the development of spot groups is influenced considerably by the neighboring older active regions depending on their age and their relative position.

A possible interpretation of this clustering of solar activity would be that all active regions of a complex originate from a persistent subsurface source rotating with a fixed period. Another possibility would be that magnetic regions primarily appearing at random generate conditions favorable for the repeated emergence of strong fields. Such a preconditioning may also be effective in the formation of multi-bipolar and still more complex spotgroups which appear as a juxtaposition or interpenetration of several groups. Leighton (1969) recently explained the tendency for new spotgroups to appear west of older spotgroups as well as the apparent persistence of preferred longitudes in terms of his comprehensive, magneto-kinematic model of the solar cycle.

It may be mentioned here that observations indicate that even widely separated active regions may be linked magnetically: Coronal arches observed in the green coronal line at the solar limb as well as structures visible on soft X-ray photographs taken recently by the Solar Physics Group, Cambridge, Mass. appear to form connections between distant active regions. Moreover, the occurrence of correlated radio-bursts above distant active regions within several seconds strongly indicates a magnetic link between the two regions through the solar corona (Wild, 1968).

Small-scale observations, on the other hand, have shown (Bumba, 1967; Bumba *et al.*, 1968; Bappu *et al.*, 1968) that the bright K plages and (with a delay of about a day) the spots of a new active region start to appear at the boundary of adjacent supergranulation cells which are known to be concentrations of magnetic flux (Simon and Leighton, 1964). The K plage emission subsequently spreads along the boundary and gradually fills a cell or two. Longitudinal fields may have opposite polarities at opposite sides of a cell or even in juxtaposition within one boundary. Transverse fields are found inside the cells or span them as a whole forming closed magnetic regions. In the first two days the development of the magnetic field and of the corresponding plage is asymmetric in that the follower part appears and develops earlier than the leader part.

The young Hα plage usually contains in its center a few dark fibrils which span a cell and show material downflow on both ends into the magnetic cell boundary (Bruzek, 1968, 1969a; Weart and Zirin, 1969). Their appearance is probably due to the emergence of magnetic field loops. It is interesting to note that the direction of the fibrils, that is the new magnetic field, has a strong N-S-component at first appearance and assumes the normal, nearly E-W-direction only after several sudden field changes within a day or so which also transform the magnetic field from a non-potential to a potential configuration (Angle and Livingston, 1969). The transverse field component measured across a cell was found to be 250 G by Bappu *et al.* (1968).

## 3. Strong Field Phenomena

Next we come to the photospheric phenomenon dominant in active regions: the *sunspots*. They appear as a local decrease of temperature, radiation and gas pressure and are a concentration of magnetic flux to maximum fieldstrengths 1200–4000 G. Apart from details, current spot theories (e.g. Deinzer, 1965; de Jager, 1964; Chitre and Shaviv, 1967; Wilson, 1968) agree that the deficit in radiative flow in the visible spot is due to the reduction of the convective transfer of energy in the convection zone by a strong magnetic field. The spot is a rather shallow feature (of the order 1000 km) which does not emerge from below but is cooled at the top of the convection zone as soon as a strong magnetic field inhibits the convective flux. The strong spot *field*, however, *cannot* be formed near the surface because of the high electric conductivity. According to Weiss (1964) it originates in the convection zone where weak magnetic fields are concentrated by cellular convection into ropes of flux of $10^{21}$ Mx and fields of 5000 G. By their own magnetic buoyancy and by convection these fluxtubes are carried to the surface where they expand because of the lower gas pressure and form the spots. Severny (1972), however, argues that the spot field has to exist independently of the action of the convection.

The energy flux missing in a large spotgroup amounts to $4 \times 10^{29}$ erg, that is, about one third of the total flux over an active region. This missing flux has, of course, to appear somewhere else. It was suggested (Kiepenheuer, 1966b) that it may show up in the bright ring of the spots, drive the Evershed motion in the spot penumbra, appear in plages and even supply energy for flares (de Jager, 1969). According to a recent estimate of Wilson (1968) only a part of the missing flux would account for all the above phenomena. We have to conclude therefore than an essential portion of the missing energy escapes observation probably since it is diverted by convection and distributed over a very large area.

As for the spot *decay* there are three possible mechanisms under discussion: Ohmic dissipation of the magnetic field would have a decay time too long by several orders because of the high electric conductivity of the surrounding solar atmosphere even if the conductivity in the spot may be reduced by $10^{-4}$ as suggested by Schröter (1966) (discussion see Altschuler, 1967; Schmidt, 1968). Radiative relaxation as proposed by Weiss (1964) would not be effective as long as the magnetic field is strong and therefore still cooling (Schmidt, 1968). 'Erosion' of the field by the action of the supergranulation as suggested by Simon and Leighton (1964) and its dissipation by the random walk mechanism (Leighton, 1964) is the most likely mechanism.

In this context tiny regions of strong magnetic fields detected recently in active regions (Sheeley, 1966, 1967; Beckers and Schröter, 1968a; Grigorjev, 1969) may be important. They do not produce visible spots, but appear in spectral lines as 'gaps' which coincide with intergranular space and bright Ca patches. *Magnetic knots* have diameters about 1100 km, a magnetic field $\lesssim 1400$ G, a magnetic flux $8 \times 10^{18}$ Mx and a lifetime $<30$ min. Beckers and Schröter (1968a) and Abdusamatov and Krat (1969) suggested that the knots are the smallest magnetic 'quanta' existing in active regions and even in spots. It is tempting to assume that the observed magnetic knots are the pieces of the field which are chipped off from the spots by the supergranulation to make them decay ('erosion' mechanism).

A strong magnetic field feature intermediate between spots and magnetic knots are the *pores*. They have typically a diameter 2″–5″, an apparent brightness 50% of the photosphere and a magnetic field $>1400$ G (Bumba, 1967; Steshenko, 1967). Rossbach and Schröter (1970) estimated that the true minimum intensity of even the smallest pores ($<2''$) may be equal to the minimum intensity of large spots, i.e. $\sim 10\%$ of the photosphere depending on the wavelength. Pores appear in intergranular space and are frequently the pre-stage of spots. According to Simon and Weiss (1970) a flux of $10^{20}$ Mx and a radius of 1500 km are the critical values for a pore to become a spot with a penumbra.

There is growing evidence that a remarkable degree of activity exists even in the dark spot umbra. There are the umbral dots which have been investigated recently in high resolution material by Beckers and Schröter (1968b). They were found to have diameters 150–200 km, about photospheric brightness, upward motion compared to their surroundings and a lifetime about 1500 sec. They appear to be an important structural element of the spot umbra and may account for certain magnetic field anomalies. Another active phenomenon in spot umbrae detected quite recently by Beckers and Tallant (1969) are the 'umbral flashes'. They are small, bright, short-lived moving knots observed in the H and K lines and the infrared triplet of ionized calcium; they are not visible in the hydrogen Balmer lines. The umbral flashes have diameters up to 2000 km, a typical lifetime of 50 sec with a tendency to repeat every 145 sec. They move towards the penumbra 40 km/sec and show a Doppler shift 6 km/sec. The exciting agent seems to have an outward velocity about 90 km/sec. It is suggested that they are produced by a magneto-acoustic wave arising from lower levels in the umbra. It is not known if they are related to the umbral dots.

Observations in the solar chromosphere, i.e. in strong chromospheric lines such as Hα and Ca II K show only a small portion of all spots still as dark features (mainly the large spots). The brightness of the chromosphere above the other spots finds two explanations: de Jager (1966) pointed out that – in spite of the inhibition of the convective energy transport in spots – there is still turbulent motion inside the spots which may produce mechanical flux sufficient to heat the spot chromosphere. Marik (1967) suggested that the heating is produced by MHD waves arising from the spot and transformed into shocks. – On the other hand, it is observed that chromospheric Hα structures frequently enter the spot from outside and cover it completely. In the same way a Hα or K plage covering a spot may also be the extension of a plage originating outside the spot.

## 4. Moderate Magnetic Field Phenomena (Active Region)

The whole active region – throughout all its levels (photospheric, chromospheric and coronal) – is permeated by moderate magnetic fields, which are partly extensions of the spot fields, partly originate outside spots (in plages). The gross configuration of the field is bipolar with large regions of opposite longitudinal fields divided by the 'neutral' or 'dividing' line or line of inversion. On a small scale the field configuration may be very complex with many 'inversions', inclusions, magnetic hills etc.

The magnetic fields – strong as well as moderate – have two main effects on the structure of the higher parts of the active region: (a) the alignment of structures in the chromosphere (most conspicuous in H$\alpha$) and the formation of loop and ray structures extending high into the corona; (b) the formation of bright plages and their coronal extensions which usually are called enhancements or condensations.

(a) *Chromospheric structures.* In the H$\alpha$ chromosphere of an active region we observe a large number of elongated filamentary structures (fibrils) which have been studied by a number of authors on high resolution photographs in recent years (Kiepenheuer, 1966a; Ramsey *et al.*, 1967, 1968; Loughhead, 1968; Nolan *et al.*, 1970; Banos and Macris, 1970). They follow the direction of the measured photospheric magnetic fields as has been established by a comparison of magnetograms and H$\alpha$ filtergrams (Tsap, 1964, 1965; Smith, 1966). The elongated – sometimes curvilinear – structures form since heat conduction as well as mass motion takes place virtually only along the magnetic field. Elevated phenomena such as loops or surges extending into the corona but observed in projection on the solar disk will, of course, in general not follow the direction of the photospheric field, they would rather indicate the field direction in their own level. This apparently offers a simple means to study the configuration of the magnetic field in the chromosphere ('chromospheric magnetograph' of Veeder and Zirin, 1970) and in the solar corona.

The only chromospheric phenomenon which does not seem to follow the direction of the magnetic field at all is the stable *filament* or *prominence*. They occupy the dividing line between the regions of opposite polarity in the active region and therefore are assumed to be directed *perpendicular* to the field connecting the two polarities. However, there is growing evidence that magnetic fields run *along* the axis of the filament: filament structures and motions are parallel to the axis, fibrils in the neighbourhood of the filament run parallel to it, the filament may terminate in regions of opposite polarities as e.g. pointed out by Smith (1968). Ioshpa (1968) and Nakagawa and Malville (1969) suggest therefore that the filaments have a two component field: an outer field perpendicular to the filament and supporting it against gravity as indicated by the underlying photospheric fields; an inside field along the filament as indicated by structures and motions as well as field measurements in prominences. Anzer and Tandberg-Hanssen (1970) recently presented a model with two magnetic fields, one produced by photospheric currents, the other by currents along the axis of the prominence.

(b) The bright *plage* and the *coronal condensation* are the dominant active region phenomena in the higher levels of the solar atmosphere. They are the source of a large variety of enhanced emissions over the whole spectrum from radio to X-ray wavelengths and therefore are most important in solar-terrestrial physics.

Plages are closely associated with moderate magnetic fields as was established by Howard (1967), and Bumba and Godoli (1968). In particular, a close correlation was found to exist between the K plage and longitudinal fields of 20–200 G. The outline of the K plage follows neatly the isogauss contour at 20–40 G and emission details are associated with inversions, magnetic inclusions and magnetic hills. In optical wavelengths plages are most conspicuous in chromospheric lines such as the H and K lines of Ca II and the low Balmer lines of hydrogen. In the EUV range line emission in active regions appears enhanced in lines originating in the upper chromosphere (e.g. H Ly$\alpha$ 1216 Å and He II 304 Å, at a temperature of the order $10^4$ K) as well as in lines of the chromosphere-corona transition region (which are lines of moderately ionized elements at $T \approx 10^5$ K) and in coronal lines of highly ionized ions such as Fe XV 284 Å and Fe XVI 335 and 361 Å at several $10^6$ K, which are hardly observed outside active regions. There is also enhanced radio emission from cm to m wavelengths as well as an enhanced X-ray spectrum which will be discussed later.

The relative enhancement of the emission compared to the undisturbed background, i.e. the contrast of the plage, increases with the stage of ionization (or, what is equivalent, with height or temperature): the photospheric facula has an excess brightness of 60%, the chromospheric plage is twice, the K plage three times (Sheeley, 1967), the O VI line 1032 Å eight times and the coronal line Fe XVI 361 Å about 40 times enhanced (Noyes *et al.*, 1970; Reeves *et al.*, 1970). The different emission plages coincide to a high degree on the solar disk (see e.g. Tousey *et al.*, 1968). Close to the limb, however, the correlation deteriorates insofar as the emission of the high ionization stages becomes relatively brighter and more extended in height (up to 60000 km) due to their origin in the corona.

On the solar limb enhancements in optical coronal lines (Fe X 6374 Å, Fe XIV 5303 Å, Ca XV 5694 Å) as well as a brightening in the white light corona are observed above plage regions. These enhancements resolve in a number of open or closed loops and rays whose shape is determined by the coronal magnetic field. Such structures exist to heights of 0.6 to 1.0 solar radius above the photosphere where the influence of the solar wind flow becomes dominant (i.e. $H^2/8\pi < \frac{1}{2}\varrho v^2$, $v$ = solar wind velocity) (Newkirk *et al.*, 1968). Then fields are dragged out into extended open streamers with still enhanced density and with neutral sheets along the axes (Pneuman, 1968, 1969; Sturrock and Smith, 1968). Coronal structures are, in general, rather stable features without conspicuous changes through many hours. Sometimes, however, fast expansions and motions are observed in association with flare occurrence (Bruzek and De Mastus, 1970).

The enhanced emissions and the increased brightness in white light are evidence for increased density and temperature. The density in coronal condensations has been estimated from their intensity in white light which is due to electron scattering of photospheric light. Measurements during solar eclipses as well as K-coronameter measurements without eclipse have been used by a number of investigators (e.g. Waldmeier, 1963; Newkirk, 1967; Bohlin and Simon, 1969; Leblanc *et al.*, 1970). Maximum electron densities in the center of the condensation, about 3 to 5 times the normal value in the undisturbed corona – i.e. several $10^9$ cm$^{-3}$ – have been found. In exceptional cases strong local condensations probably related to flare occurrence were detected with densities $\geqslant 10$ times normal values (max. $n_e = 8 \times 10^9$ cm$^{-3}$) (Saito and Billings, 1964; Dollfus, 1968; Newkirk *et al.*, 1969; Zirin, 1970).

Temperatures of coronal enhancements derived from line profiles of optical coronal lines and from the existence and ratio of different ionization stages emitting visible and EUV lines range from 2 to $4 \times 10^6$ K.

There is strong evidence for temperature and density inhomogeneities. In a recent investigation including optical coronal line, white light and 9.1 cm radio observations Leroy *et al.* (1968) found that in the early phase of evolution all radiations appear to originate in a core of high electron density and inhomogeneous temperature which is surrounded by a halo of green line emission at maximum development of the active region (spot group). In the decay phase this core disappears and density and temperature become homogeneous.

Electron densities for the entire lifetime of an active region have been derived for the first time by Dollfus (1968) and Newkirk *et al.* (1969) from K coronameter measurements. They found a fast increase of density followed by a slow spreading out and decay. Neupert (1968) used EUV emission lines of highly ionized iron (Fe XIII–XVI) and radio emission in order to study the evolution of the corona above active regions. He found the highest observed ionization stage (Fe XVI) to reach maximum first at maximum spot development. The same applies for the decimetric radio emission. The lower ionization stages of Fe had their maxima later in the decay phase of the active region. Centimetric radio enhancement exists only in young (spotted) active regions while decimeter and meter wave emissions are still enhanced above very old plage regions (Leblanc, 1969). The interpretation is that the active coronal material reaches temperatures up to $4 \times 10^6$ K and electron densities 10 times the quiet corona values at spot maximum and subsides to $2.5 \times 10^6$ K and 2–3 times normal electron densities in the decay phase.

If we turn now to the X-rays we note that the soft X-ray (1–100 Å) emission region coincides well with the K plage (Pounds *et al.*, 1968) and also conforms rather well with the shape and structure of H$\alpha$ plages (Vaiana *et al.*, 1968). Close to the limb, X-ray emission, however, shows various structures extending into heights $> 10^5$ km which are not visible in H$\alpha$. According to Pounds *et al.* (1968) 50% of the X-ray flux are emitted by a small core with 10 times normal electron density.

X-ray emission from active regions is not only enhanced but also hardened, that is, shorter wavelengths are enhanced more than longer waves. The contrast between active region and quiet sun increases considerably below 25 Å and is highest in the 1–8 Å band (Michard and Ribes, 1968). This hardening is due to increased temperature. From a study of line emission below 22 Å originating in high ionization stages (Fe XVII/XVIII, Ni XIX, Si XIV) Pounds *et al.* (1968) derived temperatures $3 \times 10^6$ to $10^7$ K.

An important feature of soft X-ray emission from active regions are fluctuations of 10% or more on a time scale of tens of minutes under non-flare conditions detected by Neupert *et al.* (1969) and Teske (1969). Sometimes these fluctuations can be identified with Hα plage fluctuations or with limb events, but often no clearly associated Hα counterpart was found. Fritz *et al.* (1967) detected substantial changes in spectral lines $\lambda < 25$ Å occurring in an interval of seconds again without visible flare activity. Apparently, X-ray emission from active regions is – at least in part – a discontinuous process. Non-thermal (microflare) processes are indicated by the observation of Acton and Fisher (1968) that even under quiet conditions radiation between 8 and 28 keV is emitted from active regions which may be bremsstrahlung from non-thermal energetic electrons.

The formation of the plage and active coronal region can be understood qualitatively as a twofold effect of the magnetic field: The mechanical flux arising in the upper part of the convection zone which heats the normal chromosphere and corona is enhanced in the moderate fields of the active region by about a factor 10 (see Kuperus, 1965, 1969, 1972). This enhanced flux is primarily responsible for the additional heating of the plage and the active corona. Its effect is increased further since the magneto-acoustic energy flux is channelled along the magnetic field and therefore will be trapped in the closed field of a bipolar region where it is either dissipated, heating the plasma or may be stored as turbulent energy (Pneuman, 1967). Extra heat will be conducted back along the closed field lines into the chromosphere and provide additional energy for the plage.

The higher temperature of the plage and the coronal condensation implies also higher density as was pointed out by Unsöld (1970). The closed magnetic field in the active region is also responsible for the stability of the coronal condensation: since the inside gas pressure is higher than outside the condensation would be unstable unless it is confined by a (stronger) closed field. Additional supply of energy may lead to expansion and to the formation of open streamers (Kuperus, 1969).

## 5. Solar Flares

The phenomenon which is most outstanding in active regions and most important in solar-terrestrial physics is the solar flare. Flares are localized and short-lived enhancements of radiation over a wide spectral range running from decameter radiowaves to energetic, hard X-rays ($E \gtrsim 100$ keV). There occur impulsive X-ray and concurrent radiobursts which are indicative of energetic electrons. High energy electrons as well as protons are detected in interplanetary space following flare appearance, and powerful mass motions and shockwaves are flare associated phenomena.

If we compare the relative amount of all types of energies expended during flares (e.g. Bruzek, 1969b), we find that the share of the particle energy is of the same order as the amount of radiative energy or even one order larger. Moreover, the first observable energetic flare phenomenon are radiations excited by fast electrons. Thus the flare appears to be essentially and primarily a particle event.

However, not every flare produces all types of energy; in particular, there are many flares which show only some of the particle phenomena or none at all. Lin (1969) for instance, found that flares may be classified as: (1) proton flares (with energetic protons, 40 keV electrons and MeV electrons); (2) electron flares (soft electrons only); (3) flares without any energetic particles. Roughly, we may distinguish between particle flares (1 and 2) and non-particle flares. Non-particle flares would be a thermal phenomenon just from the beginning, the excess heat being directly provided by the primary flare process.

Particle flares, on the other hand, may be understood as a rapid acceleration of a large number of particles to high energies which partly escape into space, partly remain trapped and supply the energy for the observed radiations (as bremsstrahlung, by collisions, by heating). Recent observations showing neatly the development of the energetic phenomena in the first phases of a particle flare support this concept of the particle flare.

The very first energetic phenomenon of many flares of all sizes (subflare to 3B flare) is an impulsive burst of hard X-rays ($E = 10$ to 150 keV) which rises to peak intensity within 60 sec and fades (exponentially) within 340 sec (average values). Rise and decay is fastest for the highest eneigies, that is, the spectrum softens rapidly during the development of the burst (Arnoldy *et al.*, 1968; Kane and Winckler, 1969). A few minutes after the hard X-ray peaks, lines of the highest ionization stage (usually Fe XXV 1.87 Å which requires 2 to $4 \times 10^7$ K) appear followed by lower stages such as Fe XXII (11.8 Å) and Fe XVII within 4 and 10–12 min respectively (Neupert, 1969). An analogous softening is observed in the soft X-ray continuum burst after an initial enhancement and hardening: the shortest wavelengths (0.5 to 3 Å band) reach maximum intensity about one minute after the hard X-ray peaks. The softer radiations 2 to 8 Å and 8 to 20 Å follow after several minutes. The emission may be enhanced by three orders in the 0.5 to 3 Å band and two orders in the 2 to 8, and 8 to 20 Å bands (Kreplin *et al.*, 1969). The emission decays the more slowly the longer the wavelength; 44 to 60 Å radiation enhancements may even outlive the H$\alpha$ flare emission.

As for the interpretation of these observations, there seems to be almost general agreement that the impulsive bursts (both hard X-ray and microwave bursts) are generated by electrons in the energy range $\gtrsim 100$ keV: the X-rays as non-thermal bremsstrahlung in the ambient gas, the microwave bursts as gyro-synchrotron radiation (e.g. Holt and Ramaty, 1969). The energetic electrons will lose energy however mainly by collisions and will be rapidly thermalized (Neupert, 1969; Teske and Thomas, 1969). Therefore, the hard non-thermal X-ray bursts soften and become thermal in character. On the other hand, the original energetic electrons entering cooler and denser regions (condensation and the chromosphere) will heat and ionize the gas and thermal bremsstrahlung and highly ionized lines will appear. During the following continuous relaxation of the electrons the region gradually cools, the bremsstrahlung continuum will soften and the ions recombine cascading through successively lower stages emitting corresponding lines and recombination continua.

About $10^{36}$ to $10^{38}$ electrons with energies of the order 100 keV have to be provided within the risetime of the hard X-ray burst in order to account for the observed hard radiation. The total energy of these electrons, $2\times10^{29}$ to $2\times10^{31}$ erg is of the order of the total radiation energy of the associated flare, and therefore could actually account for it.

A number of possible mechanisms for the particle acceleration (and for the primary flare process in general) have been proposed in recent years (Petschek, 1964; Sakurai, 1966; Sturrock, 1966, 1968; Syrovatskii, 1969; further references see Wentzel, 1964). Petschek's (1964) mechanism in the modification of Sturrock (1968) well accounts for a number of energetic flare phenomena. Annihilation (reconnection) of magnetic field lines would provide the energy and both eject particles (electrons) along the neutral sheet of a corona streamer out into the corona and space and also push electrons along the closed magnetic loops downwards into regions of higher density and stronger magnetic field. The flare energy stored in the magnetic field may be derived from the non-thermal flux which provides energy for the active region and is trapped in the closed magnetic field (see p. 108). $10^{33}$ erg could be made available within one day (Pneuman, 1967; Sturrock, 1968).

## References

Abdusamatov, H. I. and Krat, V. A.: 1969, *Solar Phys.* **9**, 420.
Acton, L. W. and Fisher, P. C.: 1968, in K. O. Kiepenheuer (ed.), 'Structure and Development of Solar Active Regions', *IAU Symp.* **35**, 432.
Altschuler, M. D.: 1967, *Solar Phys.* **1**, 377.
Angle, K. L. and Livingston, W. C.: 1969, *BAAS* **1**, 271.
Anzer, U. and Tandberg-Hanssen, E.: 1970, *Solar Phys.* **11**, 61.
Arnoldy, R. L., Kane, S. R. and Winckler, J. R.: 1968, in K. O. Kiepenheuer (ed.), 'Structure and Development of Solar Active Regions', *IAU Symp.* **35**, 484.
Banos, G. J. and Macris, C. J.: 1970, *Solar Phys.* **12**, 106.
Bappu, M. K. V., Grigorjev, V. M. and Stepanov, V. E.: 1968, *Solar Phys.* **4**, 409.
Becker, U.: 1955, *Z. Astrophys.* **37**, 47.
Beckers, J. M. and Schröter, E. H.: 1968a, *Solar Phys.* **4**, 142.
Beckers, J. M. and Schröter, E. H.: 1968b, *Solar Phys.* **4**, 303.
Beckers, J. M. and Schröter, E. H.: 1969, *Solar Phys.* **10**, 384.
Beckers, J. M. and Tallant, P. E.: 1969, *Solar Phys.* **7**, 351.
Bohlin, J. D. and Simon, M.: 1969, *Solar Phys.* **9**, 183.
Bruzek, A.: 1968, in K. O. Kiepenheuer (ed.), 'Structure and Development of Solar Active Regions', *IAU Symp.* **35**, 293.
Bruzek, A.: 1969a, *Solar Phys.* **8**, 29.
Bruzek, A.: 1969b, *Plasma Instabilities in Astrophysics*, New York, p. 71.
Bruzek, A. and De Mastus, H. L.: 1970, *Solar Phys.* **12**, 447.
Bumba, V.: 1967a, in P. A. Sturrock (ed.), *Plasma Astrophysics*, Academic Press, New York, p. 77.
Bumba, V.: 1967b, *Solar Phys.* **1**, 371.
Bumba, V.: 1972, this volume, p. 21.
Bumba, V. and Godoli, G.: 1968, in K. O. Kiepenheuer (ed.), 'Structure and Development of Solar Active Regions', *IAU Symp.* **35**, 338.
Bumba, V. and Howard, R.: 1965, *Astrophys. J.* **141**, 1492, 1502.
Bumba, V. *et al.*: 1968, in K. O. Kiepenheuer (ed.), 'Structure and Development of Solar Active Regions', *IAU Symp.* **35**, 56.
Chitre, S. M. and Shaviv, G.: 1967, *Solar Phys.* **2**, 150.
Cline, T. L., Holt, S. S., and Hones, E. W., Jr.: 1969, *Ann. IQSY* **3**, 193.
Deinzer, W.: 1965, *Astrophys. J.* **141**, 548.

De Jager, C.: 1964, *Bull. Astron. Inst. Neth.* **17**, 253.
De Jager, C.: 1966, in G. Righini (ed.), 'Proceedings of the Meeting on Sunspots', G. Barbèra, Firenze, p. 242.
De Jager, C.: 1969, in C. de Jager and Z. Švestka (eds.), 'Solar Flares and Space Research', *Cospar Symposium*, North-Holland Publ. Co., Amsterdam, p. 1.
Dodson, H. W. and Hedeman, R. E.: 1968, in K. O. Kiepenheuer (ed.), 'Structure and Development of Solar Active Regions', *IAU Symp.* **35**, 56.
Dollfus, A.: 1968, in K. O. Kiepenheuer (ed.), 'Structure and Development of Solar Active Regions', *IAU Symp.* **35**, 359.
Fritz, G. *et al.*: 1967, *Astrophys. J.* **148**, L133.
Grigorjev, V. M.: 1969, *Solar Phys.* **6**, 67.
Haurwitz, M.: 1968, *Astrophys. J.* **151**, 351.
Holt, S. S. and Ramaty, R.: 1969, *Solar Phys.* **8**, 119.
Howard, R.: 1967, *Ann. Rev. Astron. Astrophys.* **5**, 1.
Ioshpa, B. A.: 1968, in K. O. Kiepenheuer (ed.), 'Structure and Development of Solar Active Regions', *IAU Symp.* **35**, 261.
Kane, S. R. and Winckler, J. R.: 1969, *Solar Phys.* **6**, 304.
Kiepenheuer, K. O.: 1966a, *DFG Forsch. Ber.* **12**, 11.
Kiepenheuer, K. O.: 1966b, in G. Righini (ed.), 'Proceedings of the Meeting on Sunspots', G. Barbèra, Firenze, p. 26.
Kreplin, R. W. *et al.*: 1969, in C. de Jager and Z. Švestka (eds.), 'Solar Flares and Space Research', *XIth Cospar Symposium*, North-Holland Publ. Co., Amsterdam, p. 121.
Kuperus, M.: 1965, Thesis, Utrecht.
Kuperus, M.: 1969, *Space Sci. Rev.* **9**, 713.
Kuperus, M.: 1972, this volume, p. 9.
Leblanc, Y.: 1969, *Astron. Astrophys.* **4**, 315.
Leblanc, Y., Leroy, J. L. and Poulain, P.: 1970, *Astron. Astrophys.* **5**, 391.
Leighton, R. B.: 1964, *Astrophys. J.* **140**, 1547.
Leighton, R. B.: 1969, *Astrophys. J.* **156**, 1.
Leroy, J. L., Rösch, J. and Trellis, M.: 1968, in K. O. Kiepenheuer (ed.), 'Structure and Development of Solar Active Regions', *IAU Symp.* **35**, 346.
Lin, R. P.: 1969, *BAAS* **1**, 284.
Losh, H. M.: 1939, *Publ. Observ. Michigan* **7**, 127.
Loughhead, R. E.: 1968, *Solar Phys.* **5**, 489.
Marik, M.: 1967, *Soviet. Astron.* **11**, 264.
Martres, J. M.: 1968, in K. O. Kiepenheuer (ed.), 'Structure and Development of Solar Active Regions', *IAU Symp.* **35**, 25.
Martres, J. M.: 1970, *Solar Phys.* **11**, 258.
Martres, J. M. and Michard, R.: 1965, *Compt. Rend. Acad. Sci. Paris* **261**, 4336.
Michard, R. and Ribes, E.: 1968, in K. O. Kiepenheuer (ed.), 'Structure and Development of Solar Active Regions', *IAU Symp.* **35**, 420.
Nakagawa, Y. and Malville, J. M.: 1969, *Solar Phys.* **9**, 102.
Neupert, W. M.: 1968, in K. O. Kiepenheuer (ed.), 'Structure and Development of Solar Active Regions', *IAU Symp.* **35**, 404.
Neupert, W. M., White, W. A., Gates, W. J., and Swartz, M.: 1969, *Solar Phys.* **6**, 183.
Newkirk, G., Jr.: 1967, *Ann. Rev. Astron. Astrophys.* **5**, 213.
Newkirk, G., Jr., Altschuler, M. D., and Harvey, J.: 1968, in K. O. Kiepenheuer (ed.), 'Structure and Development of Solar Active Regions', *IAU Symp.* **35**, 379.
Newkirk, G., Jr., Hansen, R. T., and Hansen, S.: 1969, *Ann. IQSY* **3**, 49.
Nolan, B. A., Smith, S. F., and Ramsey, H. E.: 1970, *Solar Filtergrams*, Lockheed Solar Observ. Burbank.
Noyes, R. W., Withbroe, G. L., and Kirshner, R. P.: 1970, *Solar Phys.* **11**, 388.
Parks, G. K. and Winckler, J. R.: 1969, *Astrophys. J.* **155**, L117.
Petschek, H. E.: 1964, *AAS-NASA Symp.*, NASA SP-50, p. 425.
Pneuman, G. W.: 1967, *Solar Phys.* **2**, 462.
Pneuman, G. W.: 1968, *Solar Phys.* **3**, 578.
Pneuman, G. W.: 1969, *Solar Phys.* **6**, 255.

Pounds, K. A., Evans, K., and Russell, P. C.: 1968, in K. O. Kiepenheuer (ed.), 'Structure and Development of Solar Active Regions', *IAU Symp.* **35**, 431.
Ramsey, H. E., Angle, K. L., and Smith, S. F.: 1967, Lockheed Solar Observ. Rep. LR 20882.
Ramsey, H. E., Smith, S. F., and Angle, K. L.: 1968, Lockheed Solar Observ. Rep. LMSC 681495.
Reeves, E. M. *et al.*: 1970, paper presented at the Leningrad STP Symposium, May 1970.
Rossbach, M. and Schröter, E. H.: 1970, *Solar Phys.* **12**, 95.
Saito, K. and Billings, D. E.: 1964, *Astrophys. J.* **140**, 760.
Sakurai, K.: 1966, *Publ. Astron. Soc. Japan* **18**, 77.
Schmidt, H. U.: 1968, in K. O. Kiepenheuer (ed.), 'Structure and Development of Solar Active Regions', *IAU Symp.* **35**, 95.
Schröter, E. H.: 1966, *Florence Symp. on Sunspots*, p. 222.
Severny, A. B.: 1972, this volume, p. 38.
Sheeley, N.: 1966, *Astrophys. J.* **144**, 723.
Sheeley, N.: 1967, *Solar Phys.* **1**, 171.
Simon, G. and Leighton, R. B.: 1964, *Astrophys. J.* **140**, 1120.
Simon, G. and Weiss, N. O.: 1970, *Solar Phys.* **13**, 85.
Smith, S. F.: 1966, Lockheed Solar Observ. Rep. LR 20445.
Smith, S. F.: 1968, in K. O. Kiepenheuer (ed.), 'Structure and Development of Solar Active Regions', *IAU Symp.* **35**, 267.
Steshenko, N. V.: 1967, *Izv. Krymsk. Astrofiz. Observ.* **37**, 21.
Sturrock, P. A.: 1966, *Nature* **211**, 695.
Sturrock, P. A.: 1968, in K. O. Kiepenheuer (ed.), 'Structure and Development of Solar Active Regions', *IAU Symp.* **35**, 471.
Sturrock, P. A. and Smith, S. M.: 1968, *Solar Phys.* **5**, 87.
Syrovatskii, S. I.: 1969, in C. de Jager and Z. Švestka (eds.), 'Solar Flares and Space Research', *XIth Cospar Symposium*, North-Holland Publ. Co., Amsterdam, p. 346.
Teske, R. G.: 1969, *Solar Phys.* **6**, 193.
Teske, R. G. and Thomas, R. J.: 1969, *Solar Phys.* **8**, 348.
Tousey, R., Sandlin, G. D., and Purcell, J. D.: 1968, in K. O. Kiepenheuer (ed.), 'Structure and Development of Solar Active Regions', *IAU Symp.* **35**, 411.
Tsap, T. T.: 1964, *Izv. Krymsk. Astrofiz. Observ.* **31**, 200.
Tsap, T. T.: 1965, *Izv. Krymsk. Astrofiz. Observ.* **33**, 92.
Tsubaki, T.: 1967, *Publ. Astron. Soc. Japan* **19**, 96.
Unsöld, A.: 1970, *Astron. Astrophys.* **4**, 220.
Vaiana, G. S. *et al.*: 1968, *Science* **161**, 564.
Veeder, G. J. and Zirin, H.: 1970, *Solar Phys.* **12**, 391.
Waldmeier, M.: 1963, *Z. Astrophys.* **56**, 291.
Weart, S. R. and Zirin, H.: 1969, *Publ. Astron. Soc. Pac.* **81**, 270.
Weiss, N. O.: 1964, *Monthly Notices Roy. Astron. Soc.* **128**, 225.
Wentzel, D. G.: 1964, in W. N. Hess (ed.), 'The Physics of Solar Flares', NASA publication **50**, p. 397.
Wild, J. P.: 1968, *Proc. ASA* **1**, 137.
Wilson, P. R.: 1968, *Solar Phys.* **3**, 243.
Zirin, H.: 1970, *Solar Phys.* **11**, 497.

# PERMANENT SOURCES OF PARTICLE EMISSION FROM THE SUN

MONIQUE PICK

*Solar Department, Meudon Observatory, France*

**Abstract.** Protons of energies less than a few MeV are present continuously for periods of several days. During the last solar minimum, these protons might appear in interplanetary space with a recurrence of 27 days over many solar rotations. It is argued that the particles are of solar origin.

For the majority of cases, one may relate particle events with active regions on the sun, even during solar minimum. We thus reject the hypothesis of acceleration of particles in interplanetary space.

Solar corpuscular streams are confined in interplanetary magnetic sectors which can be relatively stable for several rotations. The 27-day recurrence is attributed to two different causes which will be discussed: the first is long-term storage of solar particles; the second is the successive formation of active centers in the same sector and at neighboring longitudes. These two causes will be discussed. Active centers responsible for solar corpuscular events may modify the magnetic sectors in which these particles are confined and by this mechanism the particle pattern itself may be modified. These modifications and some aspects of the magnetic sector origin will be discussed.

## 1. Introduction

It has been known for a long time, that very efficient corpuscular acceleration processes occur with chromospheric flares: when the energy of the accelerated particles is sufficiently high, they may be directly observed at ground level. With satellites and interplanetary space probes, numerous low energy particles events were observed even when the solar activity was at its minimum. Then, observational limitations become purely instrumental; the detectors are sensitive for any type of particle, solar or not, and large backgrounds can partially hide the temporal evolution of phenomena. Investigations involving very low energy particles then definitely require anti-coincidence detectors with sufficient geometry factors.

In this paper, we shall first recall very briefly the first experiments carried out: low energy protons (3 MeV at 20 MeV) were observed for the first time by Bryant *et al.* (1965). The experiment of Fan *et al.* took place at the end of 1963 (Fan *et al.*, 1966).

From these observations, some results are apparent as is shown by Figure 1:

(a) Low energy protons were continuously visible during periods of a few days. These protons did not seem to have their origin in specific solar flares.

(b) They were observed in several successive solar rotations with an approximate 27-day recurrence period.

(c) Simultaneous interplanetary magnetic field observations indicated that the low energy protons were confined inside interplanetary magnetic sectors, which appear as stable regions, corotating with the sun during several rotations.

## 2. Particle Origins

The primary question relative to the origin of the low energy particles is: were these

*Dyer (ed.), Solar Terrestrial Physics / 1970: Part I,* 61–71. 

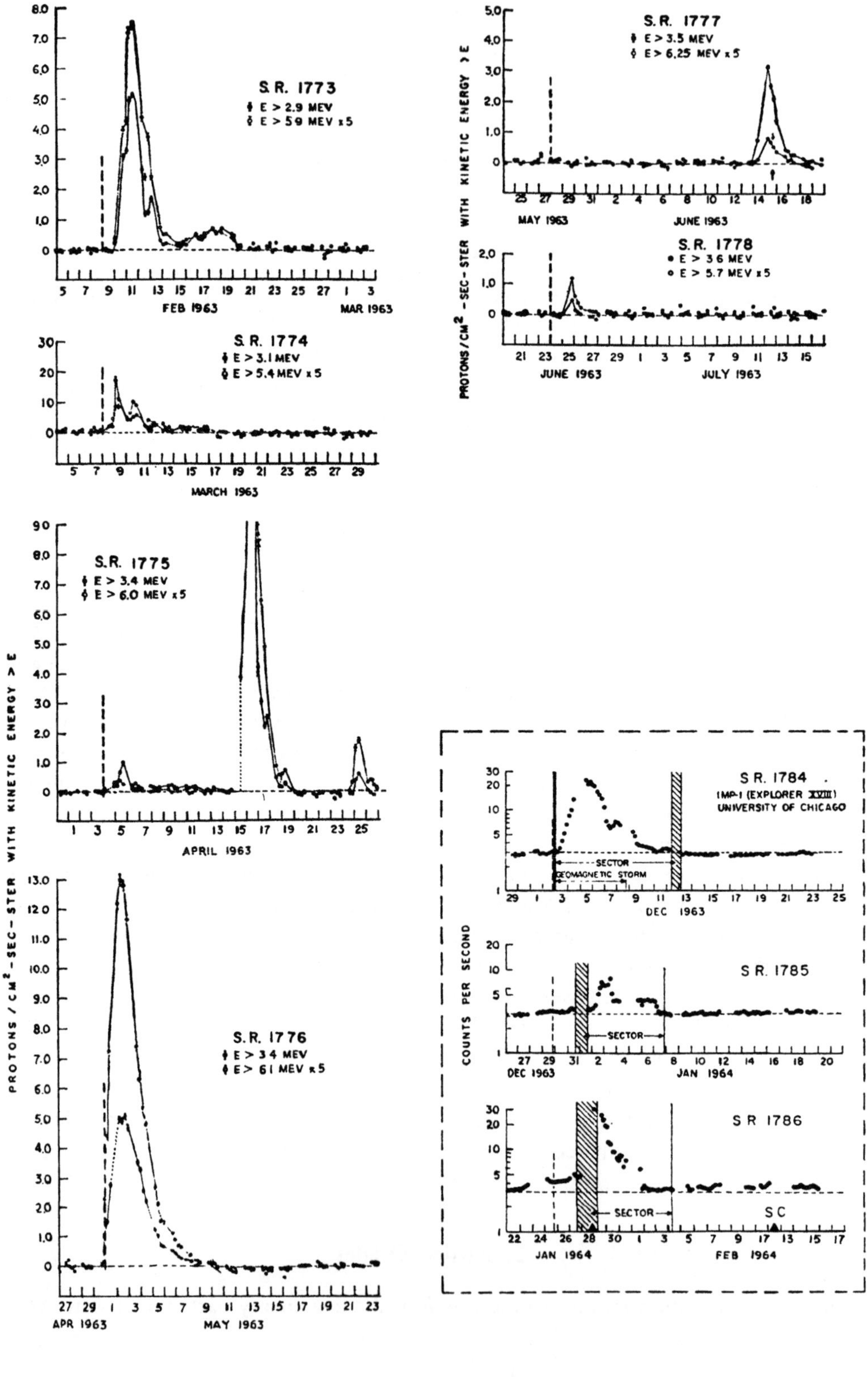
S.R. 1773
E > 2.9 MEV
E > 59 MEV x5
FEB 1963
MAR 1963
S.R. 1774
E > 3.1 MEV
E > 5.4 MEV x5
MARCH 1963
S.R. 1775
E > 3.4 MEV
E > 6.0 MEV x5
APRIL 1963
PROTONS / CM² - SEC - STER WITH KINETIC ENERGY > E
S.R. 1776
E > 3 4 MEV
E > 6 I MEV x5
APR 1963
MAY 1963
S.R. 1777
E > 3.5 MEV
E > 6.25 MEV x5
MAY 1963
JUNE 1963
S.R. 1778
E > 3 6 MEV
E > 5.7 MEV x5
JUNE 1963
JULY 1963
S R. 1784
IMP-I (EXPLORER XVIII)
UNIVERSITY OF CHICAGO
SECTOR
GEOMAGNETIC STORM
DEC 1963
COUNTS PER SECOND
S R. 1785
SECTOR
DEC 1963
JAN 1964
S R 1786
SECTOR
S C
JAN 1964
FEB 1964

particles of solar origin? If this is the case, have we or have we not flare-like acceleration process? If not, would we have to invoke another origin or another acceleration process, taking place for instance in the interplanetary medium?

On the other hand, to explain the 27-day recurrence period, is it necessary to admit a long-term storage of the particles by a magnetic field close to the sun? In this case, the trapping efficiency seems surprisingly high.

Later on, simultaneous observations by the Mariner IV space probe and several earth-orbiting satellites (IMP 1, IMP 2, IMP 3) clearly indicated that some proton events obviously associated with solar flares were also confined in the interplanetary sectors and corotated with them (O'Gallagher *et al.*, 1966; O'Gallagher, 1970).

We thus see that a certain confusion existed among events clearly associated with a flare, and those appearing with a 27-day recurrence period, both of these categories filling an interplanetary sector.

We must point out that the search for a solar origin of the events is a very delicate task during the solar minimum, because during this period every active feature, flares or active centers themselves, remains at a very small scale: it becomes very difficult to establish precise correlations with some degree of confidence.

The launching of Pioneers 6 and 7, at the beginning of the present solar cycle, has given an excellent opportunity for studying possible relationships between this low-energy component and solar activity. Particularly the telescopes from the Chicago Group on board of Pioneers 6 and 7 were able to detect protons and alpha particles of energies as low as 0.6 MeV. In these experiments, the background counting rate was reduced by using an anticoincidence shield around the telescope so that proton flux changes as small as $10^{-3}$ particles/cm$^2$/sec/ster were detected with a time resolution of 1 hour (Fan *et al.*, 1968).

The main results from these observations which are illustrated in Figure 2 are the following:

(a) Low energy protons were continuously observed for a much longer time than during solar cycle minimum. The heliographic longitude range over which enhanced proton fluxes >0.6 MeV were continuously observed in interplanetary space may be as great as 180°. As in the solar cycle minimum, these fluxes are more or less confined within interplanetary magnetic sectors.

(b) Superposed on the 0.6 MeV flux level, an occasional large intensity increase extending in energy up to more than 50 MeV for protons produced by a discrete flare was observed.

(c) It was shown that proton fluxes are associated with specific active regions on the sun within which the principal flare activity resides. This result clearly shows that large proton increases are associated with the existence on the solar surface of strong eruptive centers. The synoptic chromospheric chart displayed in Figure 3 shows

---

Fig. 1. Low energy solar cosmic rays. Compilation of the Goddard Space Flight Center: Explorer XIV (Bryant *et al.*, 1965) and University of Chicago IMP A data (Fan *et al.*, 1966). This compilation is done by Fichtel and McDonald (1967). For the last three recurrences the position of sector boundaries observed by Wilcox and Ness (1965) is indicated (Fichtel and McDonald, 1967).

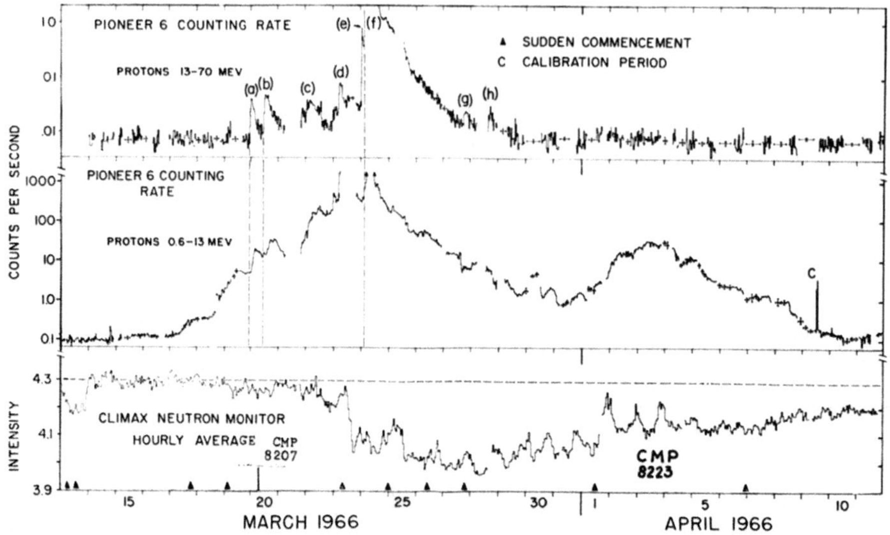


Fig. 2. Thirty-minute averages of the counting rates of protons 13–70 and 0.6–13 MeV. The enhanced flux of 0.6 to 13 MeV protons during March 15–31 is attributed to Solar Region 8207. The first evidence of enhanced flux from the following region (8223) appears on March 31 (Fan *et al.*, 1968).

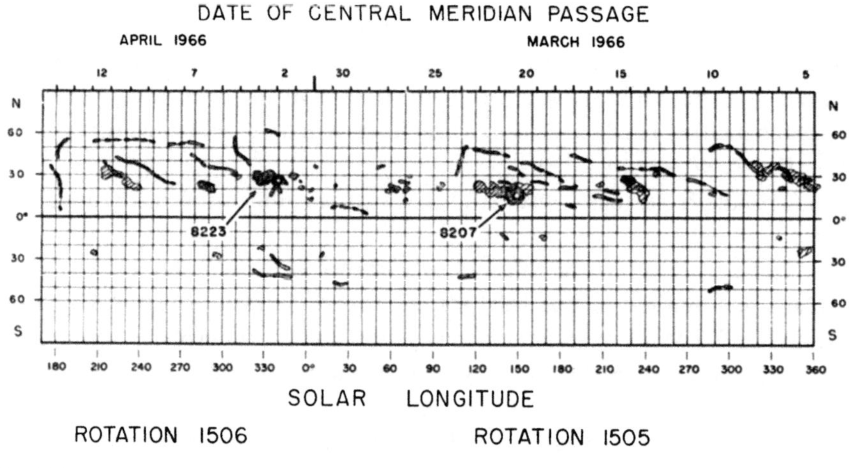


Fig. 3. Solar map showing the location of plage regions 8207 and 8223 on the sun (adapted from the synoptic charts of the solar chromosphere, Meudon Observatory, France). Rotation numbers follow the international convention. Carrington coordinates (fixes on the solar disk) are used (Fan *et al.*, 1968).

according to Fan *et al.* (1968) the two active regions chiefly responsible for proton fluxes observed for the two March 1966 periods.

The apparent positive correlation of these events with active regions led us to re-examine the recurrent events during the cycle minimum (1963–1964 period). It is noted that for almost every proton event there existed at least one solar active center which gave rise to several chromospheric flares as illustrated in Figure 4. It is also noted that in this period the majority of active centers occur on the sun, within a range of Carrington longitudes of 90°, during a period of about 15 months.

It seems that we must conclude, that whatever the period of the solar cycle, the low energy particles continuously observed during many days are in the majority of cases closely related to the presence on the solar surface of centers of activity, more or less eruptive according to the period within the cycle.

Recent analyses of particle spectra observed in the energy interval 4–80 MeV by Kinsey (1969) confirmed the fact that the sun appears as a quasi-continuous source of solar cosmic rays. This is also in agreement with results of Mustel who studied the correlation between active regions and M disturbances and concluded that practically every active region is a permanent source of particles (Mustel, 1964). Therefore it is not necessary to invoke a long-term storage of particles over many consecutive rotations. Another argument against a very long-term storage inside the interplanetary medium comes from the anisotropic distribution of these events. This anisotropy was first observed by Rao *et al.* (1967) and more recently by Frank for low energy proton fluxes with an energy range $5 < E < 50$ keV: it exists in the angular distribution in the ecliptic plane with largest intensities for particles arriving from near solar directions; the intensity ratio amounts to a factor of 2 to 3 (Frank, 1970) (Figure 5). This point is also discussed by Axford (1972) in his review paper.

## 3. Origin of the 27-Day Recurrence

We should like now to discuss the origin of the 27-day recurrence. An explanation was advanced by Fichtel and McDonald (1967) and more recently by Kinsey (1969).

The 27-day recurrence can be attributed perhaps to three different causes:

(a) The first is the successive formation of active centers at neighboring longitudes: for instance, the solar active centers responsible for the principal flare activity from December 1965 through mid-1966 were located in two principal longitude regions on the sun, approximately 180° apart.

The same conclusions can be drawn from the 1967 periods, for which some recurrent events, reported by Kinsey (1969), indicate a correlation between proton fluxes and the central meridian passage of calcium plage regions. It is seen in Figure 6 that the solar situation is changing in the same range of Carrington longitudes, rotation after rotation; at each rotation, new eruptive centers are visible.

(b) The second cause is the formation of a stable and recurrent magnetic interplanetary sector which contains the low energy protons when these exist. During the minimum of the solar cycle this sector can be relatively stable for periods of several solar rotations.

(c) The third cause concerns the conditions of propagation in the sector itself; it is closely related to the second since it concerns directly the origin of the formation of an interplanetary sector. There may exist a privileged propagation region inside the sector. Indeed, as suggested by McDonald, the recurrent events may have relatively

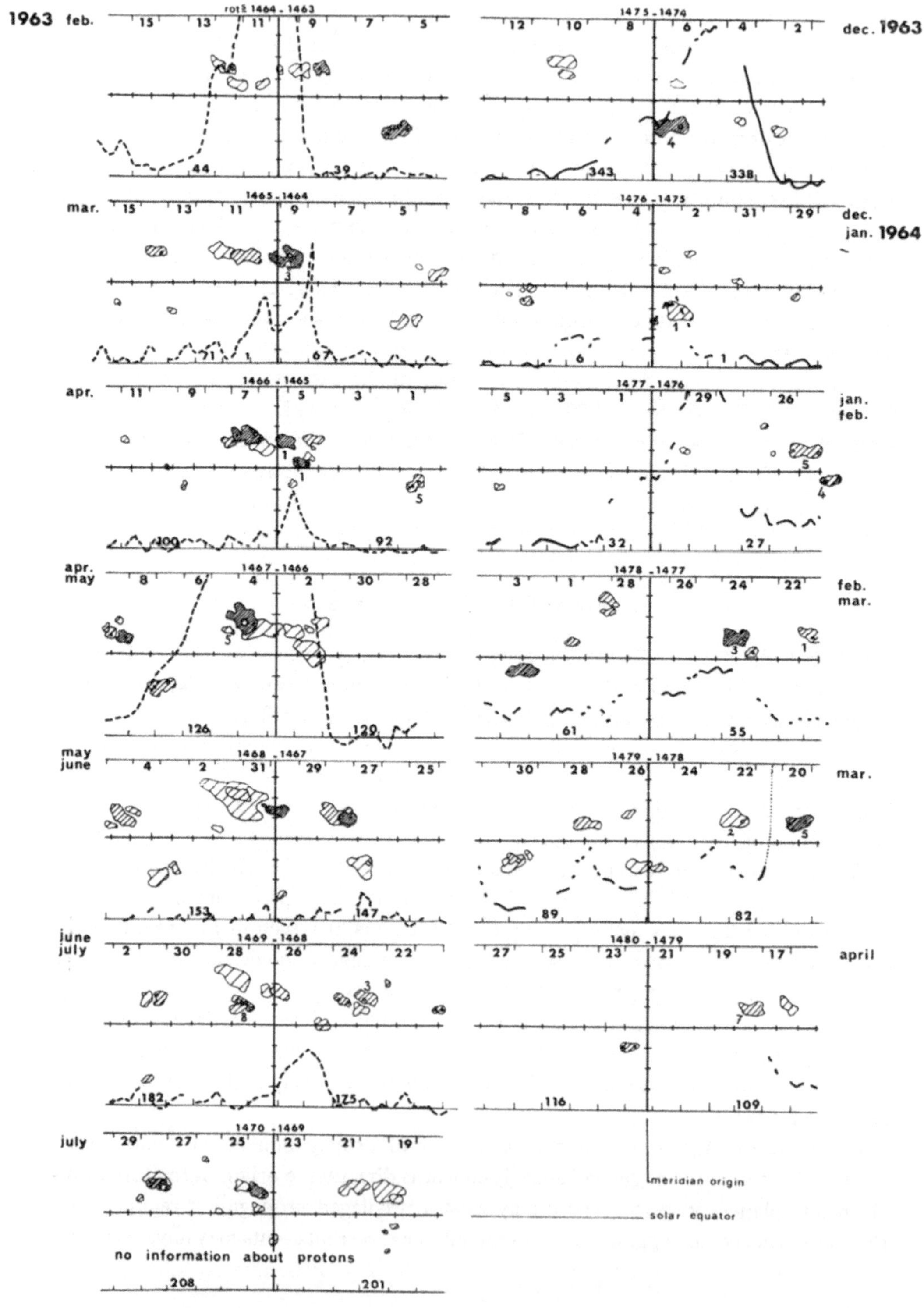
1963 feb.
rot.s 1464 - 1463
mar.
1465 - 1464
apr.
1466 - 1465
apr. may
1467 - 1466
may june
1468 - 1467
june july
1469 - 1468
july
1470 - 1469
no information about protons
dec. 1963
1475 - 1474
dec. jan. 1964
1476 - 1475
jan. feb.
1477 - 1476
feb. mar.
1478 - 1477
mar.
1479 - 1478
april
1480 - 1479
meridian origin
solar equator

characteristic temporal variation in the sector as indicated on Figure 7. McDonald notes the existence of a maximum followed by a kind of 'plateau'.

Recently, Martres *et al.* (1970) were able to show the importance of coronal streamers associated (or not) with large facular 'plages' in the forming of interplanetary sectors. These regions are clearly associated with the origin of a sector. Indeed, taking

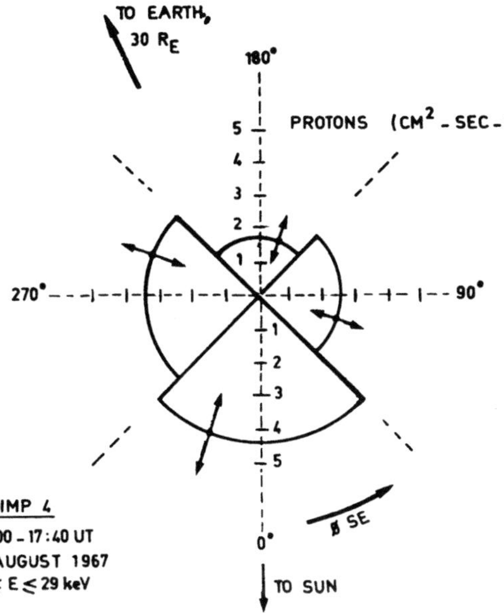


Fig. 5. Angular distribution of interplanetary proton ($11 \leqslant E \leqslant 29$ keV) intensities during the event on August 11, 1967 (Frank, 1970).

into account the sun-earth transit time (approximately 4 days), the interplanetary sector boundaries have been found to lie systematically to the west of a stable coronal streamer observable at meter wavelengths. These streamers suggest the presence of a magnetic field which is regular and practically radial and it is proposed that they are the stable source regions from which interplanetary magnetic fields emanate.

We may assume that these large coronal regions with a longitudinal extension of about 50°, where the magnetic field is uniform and radial, correspond to regions of privileged propagation for the particles. These particles could in fact be accelerated somewhere in the sector over the active regions but propagate more readily in the first 50° of the interplanetary sector, where the field is regular and radial. It is possible to find some evidence for this by re-examining certain cases of recurrence described by McDonald, and occurring during solar minimum when active centers have a rather small efficiency to produce proton enhancements; they are shown in Figure 7.

Fig. 4. Solar active centers observed during periods in which 'recurrent' protons are observed (from 'Cartes synoptiques de la chromosphère solaire', Meudon).

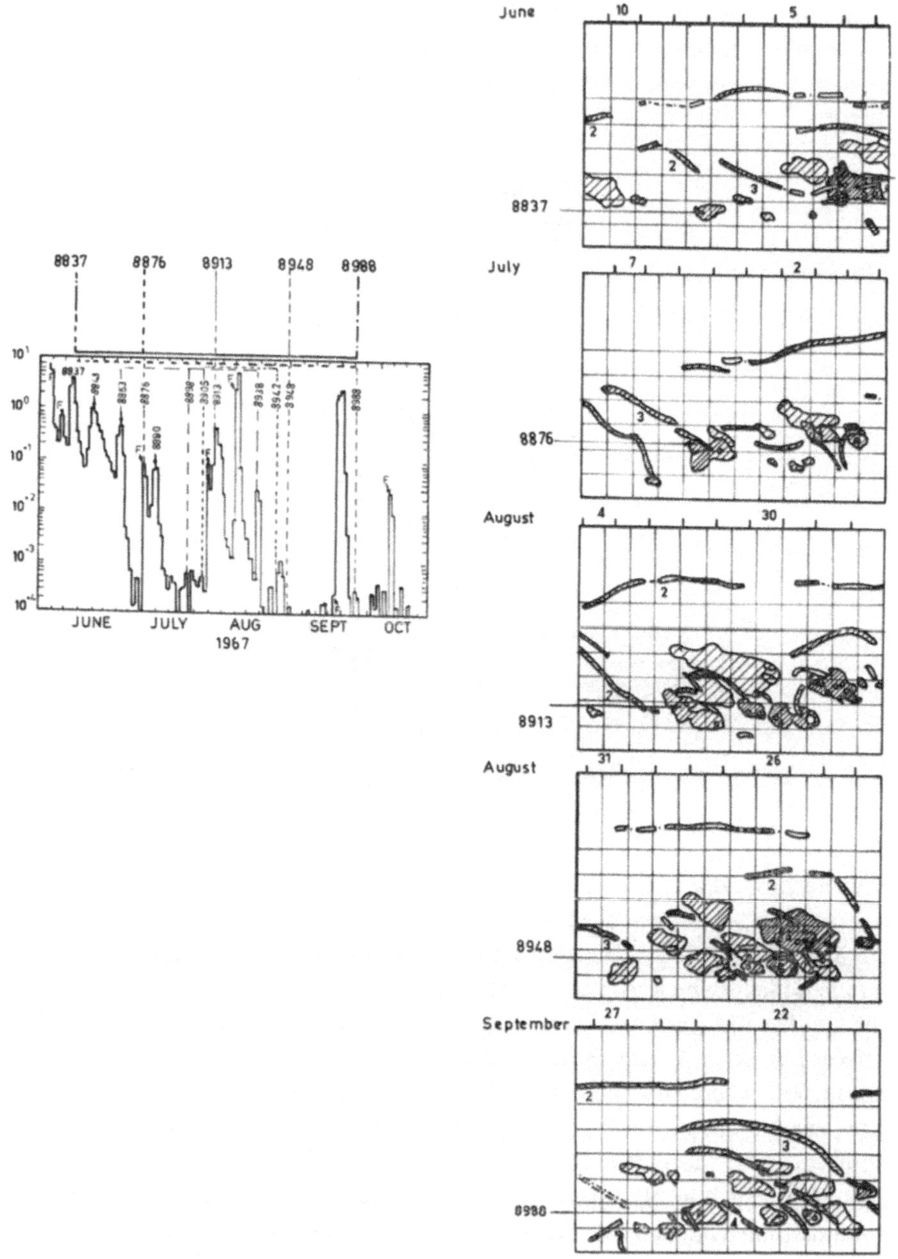


Fig. 6. Twenty-four-hour average proton fluxes at 5.2 MeV for IMP IV. The times where large confirmed solar flares occurred are marked *F*. Central meridian passage of calcium plage regions are marked by the plage region number. Recurring regions are connected by horizontal lines. Synoptic charts of the solar chromosphere for the regions concerned (Meudon Observatory) (Kinsey, 1969).

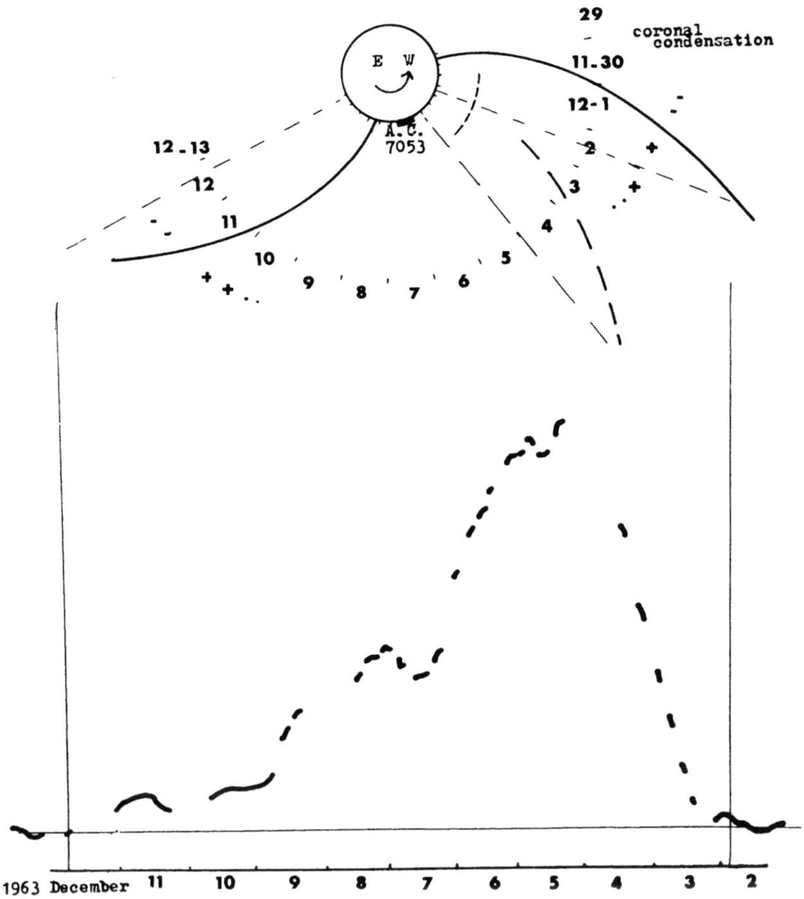


Fig. 7. Comparison between observation of an interplanetary sector in a period of solar minimum and the counting rate detecting low energy protons Imp I (Explorer XVIII). Interplanetary sector is given by Wilcox and Ness (1965).

However, when important eruptive centers exist, their position at the solar surface becomes the predominant factor for explaining temporal proton profiles.

A question which still remains open is what processes are involved in filling the sectors with particles. Several processes may combine their effects: diffusion of particles in magnetic irregularities or random walk of magnetic fields or more simply direct channeling near the sun itself. In this respect we may give some recent results showing indirectly, by radio observations, large coronal magnetic field patterns, as is schematized in Figure 8 (Lantos-Jarry, 1970), which represent possible magnetic channels.

## 4. Acceleration Processes

We should now like to discuss fairly briefly the problem of acceleration of the low energy particles visible continuously on the sun over several consecutive days. Are they associated with a series of micro-eruptions taking place in an active center? It may be suggested that this type of acceleration is connected to purely coronal instabilities which do not necessarily generate chromospheric eruptions with a flash phase. It could be very frequent: indeed, radio observations furnish us with many examples of the existence of accelerated particles not closely associated with eruptions. Hence, at very low frequencies, true type III storms can be observed continuously over periods of several days. These observations have been reported by Erickson. These measurements were performed recently at very low frequencies aboard a satellite by Alexander *et al.* (1969).

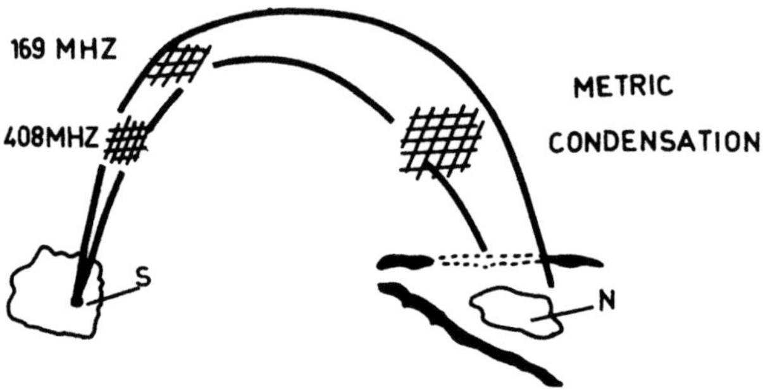


Fig. 8. An example of coronal magnetic field pattern (Lantos-Jarry, 1970).

In connection to this observation, we can recall the correlation suggested by Lin and Anderson (1967) between the presence of 40 keV electrons observed continuously in the interplanetary medium and that of metric type I storms. Sometimes these type I storms are seen as type III at very low frequencies. On the other hand, it seems that before type IV bursts, in the hour or half-hour which precedes the eruption, purely coronal activity is observed at meter wavelengths (Wild, 1968; Daigne, 1970).

## Acknowledgements

It is a pleasure to thank Prof. C. Y. Fan for a critical reading of a first draft of this paper and for many suggestions.

I am also indebted to Prof. Mustel, F. Axisa and M. J. Martres for helpful discussions.

## References

Axford, W. I.: 1972, in E. R. Dyer, (ed.), 'The Interplanetary Medium', *Solar Terrestial Physics* 1970, Part II, p. 110.
Alexander, J. K., Malitson, H. H., and Stone, R. G.: 1969, *Solar Phys.* **8**, 388.
Bryant, C. D., Cline, T. L., Desai, U. D., and McDonald, F. B.: 1963, *Phys. Rev. Letters* **11**, 144.
Bryant, C. D., Cline, T. L., Desai, U. D., and McDonald, F. B.: 1965, *Phys. Rev. Letters* **14**, 481.
Daigne, G.: 1971, in R. Howard (ed.), 'Solar Magnetic Fields', *IAU Symp.* **43**, 367.
Fan, C. Y., Gloecker, G., and Simpson, J. A.: 1966, *Proceedings of the Ninth International Conference on Cosmic Rays*, London, **1**, 105.

Fan, C. Y., Pick, M., Pyle, R., Simpson, J. A., and Smith, D. R.: 1968, *J. Geophys. Res.* **73**, 1555.
Fichtel, C. E. and McDonald, F. B.: 1967, *Ann. Rev. Astron. Astrophys.* **5**.
Frank, L. A.: 1970, *J. Geophys. Res.* **75**, 707.
Kinsey, J. H.: 1969, Ph.D. Thesis, University of Maryland.
Lantos-Jarry, M. F.: 1970, *Solar Phys.* **15**, 40.
Lin, R. P.: 1970, *Solar Phys.* **12**, 266.
Lin, R. P. and Anderson, K. A.: 1967, *Solar Phys.* **1**, 446.
Martres, M., Pick, M., and Parks, G. K.: *Solar Phys.* **15**, 48.
Mustel, E. R.: 1964, *Space Sci. Rev.* **3**, 139.
O'Gallagher, J. J.: 1970, *J. Geophys. Res.* **75**, 1163.
O'Gallagher, J. J. and Simpson, J. A.: 1966, *Phys. Rev. Letters* **16**, 1212.
Rao, U. R., McCracken, K. G., and Bukata, R. P.: 1967, *J. Geophys. Res.* **72**, 4343.
Wilcox, J. M. and Ness, N. F.: 1965, *J. Geophys. Res* **70**, 5793.
Wild, J. P.: 1968, *Proc. A.S.A.* **1** (4), 137.

# SOLAR DISCRETE PARTICLE EVENTS

Z. ŠVESTKA

*Astronomical Institute of the Czechoslovak Academy of Sciences, Ondřejov*

**Abstract.** The review paper discusses the following items in particular: Classification of particle events and the problem whether all flares are associated with particle emission of some kind; relations of the flare capability of particle emission to properties of associated active regions; differences in the size of proton events and the reasons for them; our improving knowledge of electron events, pure electron and proton events, and the possibility of two different acceleration processes on the sun; directly observed evidence of acceleration processes on the sun and some consequences deduced from these observations; the acceleration phase in the flare development.

## 1. Classification of Particle Events

In the last few years, the very sensitive detectors on board of satellites and space probes have revealed that the emission of particles from the sun is a much more frequent phenomenon than had been supposed before, and that the recorded particle events greatly differ as to their character. Table I tries to demonstrate the variety of phenomena we meet with when speaking about particle emission from the sun.

TABLE I

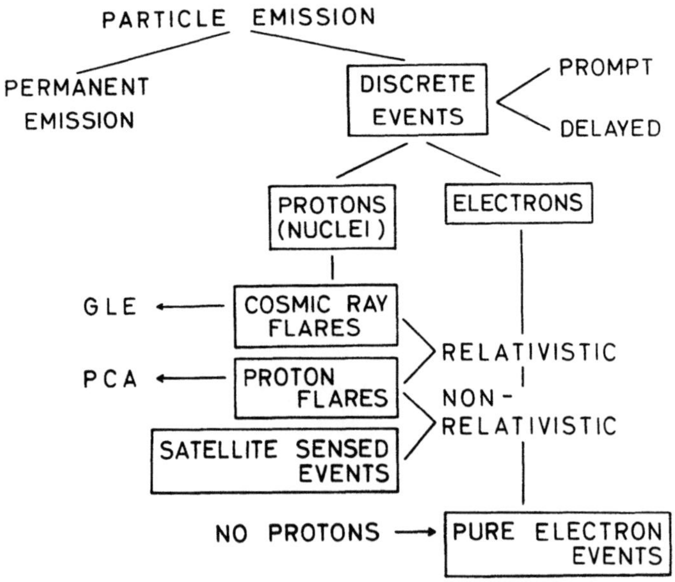


### 1.1. Permanent emission

Permanent emission of protons with energy close to and below 1 MeV is associated with some specific active regions, as we heard in the preceding lecture (Fan *et al.*, 1968; Pick, 1971). It is not yet clear whether this flux manifests a permanent accelera-

*Dyer (ed.), Solar Terrestrial Physics / 1970: Part I, 72–86.* 

tion of particles in the active region or a storage of low-energy particles in space remaining there as remnants of individual discrete particle events. In any case, however, these active regions are the seats of discrete particle events, and thus they represent regions where favourable conditions for acceleration processes have been formed (Švestka, 1970a).

## 1.2. Prompt and Delayed Events

The discrete particle events can be classified from two points of view. First, we distinguish *prompt* and *delayed* events (Figure 1). The prompt events set in tens of minutes or few hours after the particle ejection on the sun, and are directly connected with an acceleration process in the solar atmosphere. Delayed events, on the other hand, occur tens of hours after the suspected flare on the sun, and often their association with acceleration processes in the solar atmosphere is not clear (Anderson, 1969). Some of them may be associated in a general way with solar active centers as discovered by Fan *et al.* (1968). There are events which consist of both these components, and the delayed event is often associated with magnetic inhomogeneities or with shock waves in space (Figure 1).

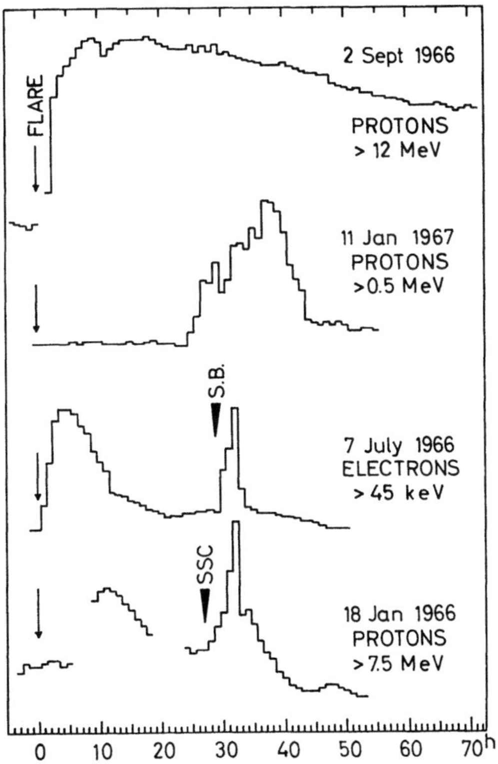


Fig. 1. Examples of prompt and delayed particle events: Sept. 2, 1966 – prompt (Lindgren, 1968); Jan. 11, 1967 – delayed (Anderson, 1969); July 7, 1966 – prompt and delayed, associated with a sector boundary (S.B.) crossing (Kahler and Lin, 1969; Švestka, 1968); Jan. 18, 1966 – prompt and delayed, associated with a sudden commencement (McCracken, 1969).

### 1.3. Prompt solar proton events

The second classification is according to the *species* of the recorded particles (Table I). These may be either atomic nuclei, among them protons as the most common sort, or electrons. Since on some occasions only one of these two components is observed, it is useful to distinguish and discuss the proton and electron solar events separately.

We shall pay most attention to prompt solar proton events. According to their size, we distinguish *cosmic-ray flares*, which cause a ground level effect (GLE) in records of neutron monitors at the sea-level, and in which there must be enough protons with energy above 500 MeV in the earth's distance from the sun. These are extremely rare phenomena, and since 1942 no more than 20 cosmic-ray flares have been recorded (Hultqvist, 1969; Dodson and Hedeman, 1969). Next we distinguish events that produce polar cap absorption (PCA) in the polar ionosphere, which means that there must be strong flux of protons with energy above 10 MeV close to the earth, even when relativistic protons are absent. Particle emissions of this type have been called *proton flare events* and about 170 such cases have been recorded since 1956 (Fritzová and Švestka, 1966; Hultqvist, 1969). And then, there are hundreds of smaller *satellite-sensed events* which never could be recognized at ground-based stations.

### 1.4. Size of a particle event

Generally, the size of an event, as it is recorded at the earth or at a spacecraft, is determined by three parameters:

The total number of accelerated particles in the flare-region, $N$; the shape of the energy spectrum, which can be characterized, e.g. by the $\gamma$ exponent in an integral power-law energy spectrum,

$$N(>E) = C \cdot E^{-\gamma};$$

and by the angular distance $\vartheta$ of the particle source on the solar disc from the root point of the magnetic-field-line which directly connects the sun with the detector.

Figure 2 shows schematically the distribution of particle sources on the solar disc, which cause strong events, weak events, and permanent emission, respectively: 84% of all strong PCA's before 1963 had their origin in flares situated between 10° east of the central solar meridian and the western limb (Švestka, 1966). On the other hand, when all events are considered, not taking into account their importance, there is no striking prevalence on the western hemisphere, and in fact, such events are recorded from flares that occur within 70° east far behind the western solar limb (Fritzová and Švestka, 1966), which approximately is the extent in longitude from where the permanent particle emission also is recorded (Fan *et al.*, 1968).

Of course, there are exceptions, probably caused by anomalous diffusion either in the solar corona or in interplanetary space, when protons succeed in reaching the earth or spacecraft even from the most remote part of the solar surface (Dodson *et al.*, 1969), or when events with quite large $\vartheta$ value become unexpectedly strong (see, e.g., Dodson and Hedeman, 1969). Examples of three such anomalous events are shown in the figure.

This concerned the parameter $\vartheta$. In order to demonstrate the role of the parameters $N$ and $\gamma$, Figure 3 shows examples of energy spectra of several selected proton events. On February 23, 1956 and September 2, 1966, $N$ was very high, and the number of protons was about the same in the energy range which determines the PCA size.

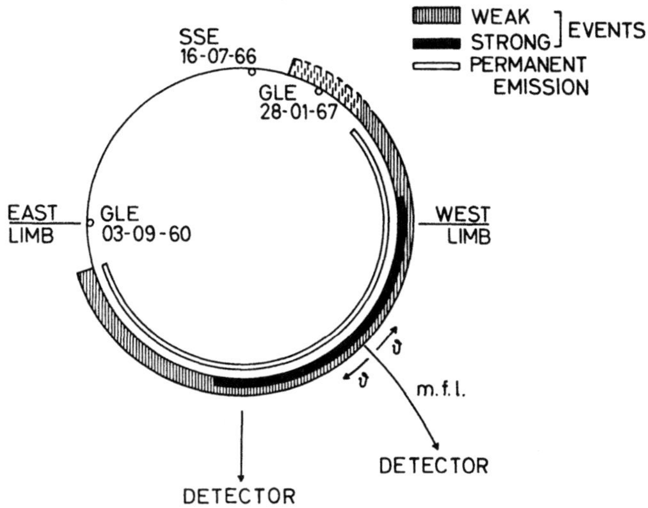


Fig. 2. Schematical drawing of the distribution of sources of strong and weak proton events on the solar surface. $\vartheta$ measures the angular distance from the magnetic field line that directly connects the sun with the detector. The figure also shows the range, within which permanent proton emission is recorded from particular active regions. Examples of three exceptional events are shown by circles and dates.

PCA's also were of about the same strength, 13 dB. But the exponent $\gamma$ for energies above 100 MeV was about twice as high in the September event as in that of February. In consequence of it there were many protons with energy exceeding 500 MeV in the February event, and an extremely strong cosmic ray increase was recorded, while protons with such high energies were absent in the September event, and no GLE was recorded at all. On the other hand, the well-known cosmic-ray flare of July 7, 1966 was characterized by a very small $N$. The $\gamma$ exponent, however, was very small, the spectrum was very hard, and a small GLE was recorded by ground-based neutron monitors, in spite of $N$ being two orders of magnitude lower than in the September event, when no GLE was observed.

### 1.5. Electron Events

Let us now discuss briefly the electron events (Table I). On rare occasions *relativistic electrons* have been recorded in space (Cline and McDonald, 1968; Datlowe *et al.*, 1969; Koechlin *et al.*, 1969), and as far as data are available, all these events were observed in association with proton flares or cosmic ray flares (Lin, 1970). Thus

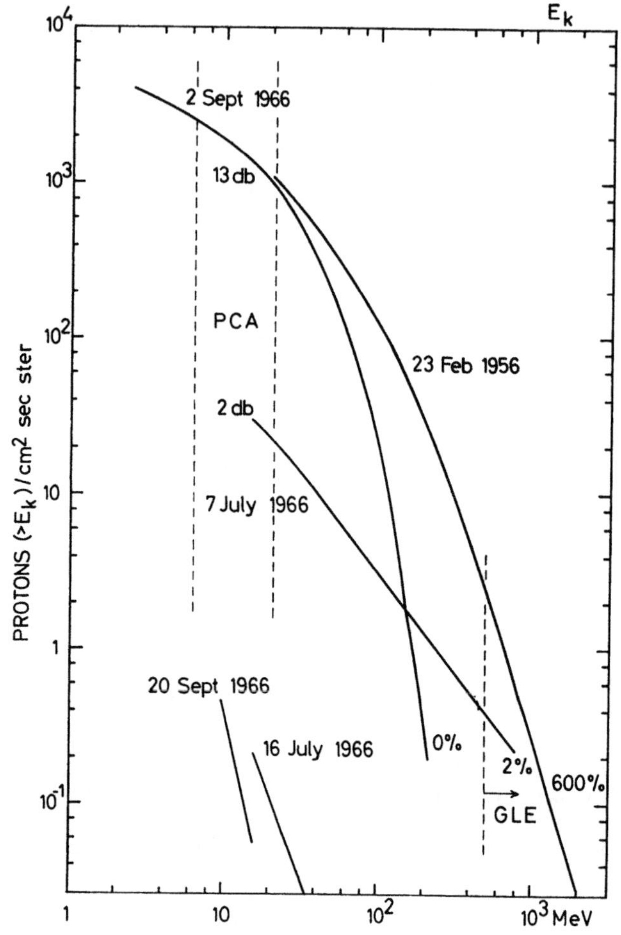


Fig. 3. Energy spectra of selected particle events showing the role of the parameters $N$ and $\gamma$ in the PCA and GLE production: 23 Feb. 1956 cosmic ray flare: $N(>20\text{ MeV}) = 1070$, $\gamma(>100\text{ MeV}) = 3.2$, both GLE and PCA strong; 7 July 1966 cosmic ray flare: $N(>20\text{ MeV}) = 23$, $\gamma(>100\text{ MeV}) = 1.3$, both GLE and PCA weak; 2 Sept. 1966 proton flare: $N(>20\text{ MeV}) = 1000$, $\gamma(>100\text{ MeV}) = 6.0$, strong PCA, no GLE. Satellite sensed events of 16 July and 20 Sept. 1966: $N(>20\text{ MeV}) \leqslant 0.1$, no effects at the earth. Data taken from Baker *et al.* (1969), Hultqvist (1969), Kinsey and McDonald (1968), McCracken (1967), Smart and Shea (1969), and Švestka (1970a).

relativistic electrons obviously are produced by the same acceleration process, which gives rise to protons of the same kinetic energy, of the order of MeV or tens of MeV. The situation, however, is quite different with *non-relativistic electrons* of energy of tens of keV, which are observed very frequently in association with proton events of all types, but often they occur without any associated proton component at all. Thus we observe very often *pure electron events* in space, which leads to the conclusion that there are two separate particle acceleration mechanisms on the sun, one of which produces low-energy electrons and the other of which produces solar protons and relativistic electrons (Lin, 1970).

On the other hand, Arnoldy *et al.* (1968) have reported on several events when only protons were recorded, without any electron component associated. For flares far on the eastern hemisphere this absence of electrons may be due to differences in propagation, because according to Lin and Anderson (1967; Lin, 1970) the electron events are even more concentrated to the western solar hemisphere than the proton events. Larger values of $\vartheta$ only occur in so-called complex events which are accompanied with a type I radio noise storm region. Nevertheless, this explanation certainly cannot be applied to at least one conspicuous event which occurred on 4 October, 1965, 30 °W from the solar central meridian, and which was anomalous from several points of view. This event was sufficiently strong in protons to produce a weak PCA effect (Hultqvist, 1969), but did not contain any electrons, did not produce any X-ray burst, and its microwave type-IV burst component was almost completely missing (Arnoldy *et al.*, 1968). Hence a striking deficiency in accelerated electrons existed not only in interplanetary space, but in the flare region itself, as well. While pure electron events are observed quite often, it is difficult to say, at the present time, how often the 'pure proton events' occur. According to Lin (1970) there were 4 such events since 1964 to the end of 1967, while 77 pure electron events were observed during the same period.

In delayed events, the most unusual feature is that the electron flux is centered several hours before the proton flux (Anderson, 1969). It looks like a corotation effect, which would show that if particles are stored in space for about twenty or more hours, a separation of electrons and protons occurs and the electrons are advanced in a direction to the west by an amount corresponding to 1 to 7 degrees with respect to the protons. But it is fairly difficult without very artificial and unrealistic assumptions, to find a reason for a drift in interplanetary space large enough to account for this rather large separation. Another observation which is difficult to explain has been made by Rao *et al.* (1969) who have found that electrons observed subsequent to a flare are only mildly anisotropic, while the protons of low energy observed at later times usually show quite a sharp anisotropy.

## 2. Particle Acceleration

There are contradictory opinions on the problem whether particle emission of some kind is a general characteristic feature of all solar flares. What I believe is that this is not true. All flares obviously are associated with a heating of a limited volume in the corona to temperatures of some millions or tens of millions degrees, if the flare is big. Thus a hot condensation is formed above the flare, but the average kinetic energy of electrons in such a condensation does not exceed a few keV. Electrons from the high-energy tail of the energy spectrum might escape into space, but their number is necessarily very low. In order to get an increased number of high-energy electrons and protons, an *impulsive process* must set in, which accelerates some fraction of particles to substantially higher energies. And this impulsive process is obviously not present in all flares.

### 2.1. Radio bursts

There are several direct manifestations of impulsive acceleration processes on the sun (Figure 4). One of them are radio type III bursts, produced by accelerated electron streams in the solar corona. Most pure electron events are associated with type III bursts, often accompanied with a type V continuum (Lin, 1970; Švestka, 1969), which indicates trapping of accelerated electrons high in the corona. Type III bursts, however, are not associated with proton events and therefore, they obviously do not represent the most efficient acceleration process which produces the proton flares. Proton flares, on the other hand, are generally associated with radio type IV bursts, i.e. strong continuous emission throughout the whole radio spectrum. This emission is due to synchrotron radiation of mildly relativistic and relativistic electrons and therefore, it is a direct evidence that an efficient acceleration process occurred in the flare region.

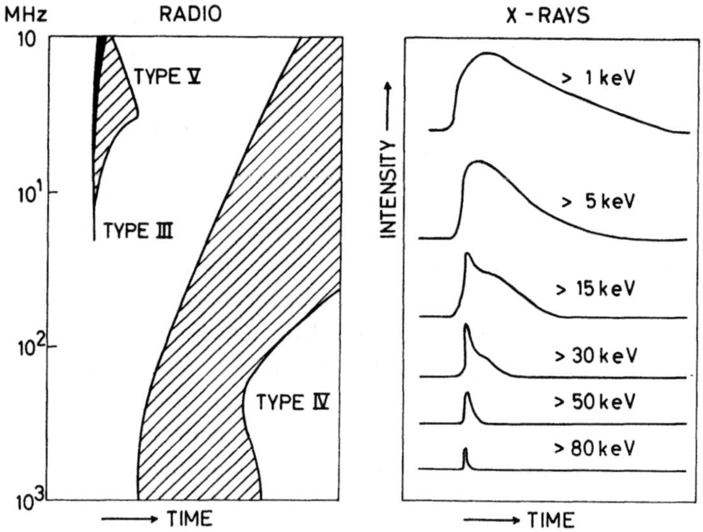


Fig. 4. Examples of direct evidence of impulsive acceleration processes on the sun.

As Castelli *et al.* (1967) have shown, strong particle events are characterized by a U-shaped type IV burst (Figure 5), with high intensity in the microwave and metric regions, and with a minimum flux between 600 and 2000 MHz. It seems that the strong flux at very high frequencies is the main factor which characterizes important particle events, and this also has been supported by Croom (1970) who has found a good correlation of particle events with intense bursts at millimeter wavelengths. In fact, strong microwave and millimeter bursts reflect a close proximity of flare regions to sunspots and covering of spots by the flaring areas (Malville and Smith, 1963), which is another characteristic feature of proton flares. On the other hand, according to Sakurai (1969), the low-frequency flux increase in the type IV burst spectrum seems

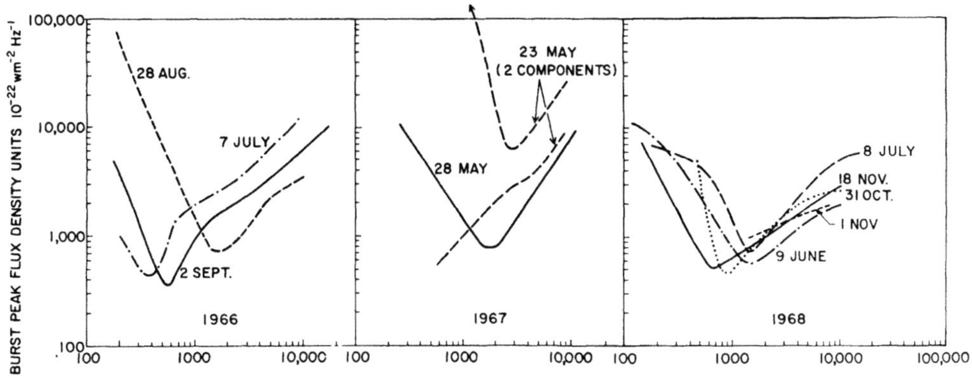


Fig. 5. U-type peak flux spectra of solar radio bursts associated with PCA's ⩾ 2.0 dB at 30 MHz in 1966–1968. (After Castelli and Aarons, 1969.)

to be characteristic for flares ejecting storm-producing blast waves. The U-shape of the burst would then be a natural consequence of the fact that strong emission of high-energy particles is generally associated with blast wave production.

### 2.2. X-RAY BURSTS

Another direct evidence of an impulsive acceleration can be found in records of X-ray bursts (Figure 4). While soft X-rays are purely thermal in their origin, this thermal component weakens with the increasing energy and it finally disappears, leaving only an impulsive short-lived X-ray burst for energies above about 50 keV. As Kane (1969) has shown, there are flares which do, and other flares which do not contain the impulsive component, which is evidence that impulsive acceleration does not occur in all flares.

### 2.3. THE ACCELERATION PHASE IN THE FLARE DEVELOPMENT

All the effects that manifest an impulsive acceleration process occur in the early, flash phase of the flare development, before the flare maximum in the Hα light (Zirin and Russo Lackner, 1969; Švestka and Simon, 1969; Švestka, 1970b). It is true that the radio and X-ray bursts give an evidence only on the acceleration of electrons to mildly relativistic energies. Since, however, there is no indication of any acceleration process later in the flare development, it is reasonable to suppose that *the whole acceleration including that of electrons and protons to energies of tens and hundreds of MeV, is accomplished in this relatively very short flash phase of the flare.*

It is easy to show that the number of relativistic electrons must be strongly suppressed due to energy losses caused by magnetic bremsstrahlung and in fact, there must be a high-energy cutoff for electrons, the cutoff energy depending on the magnetic-field strength in the region where the acceleration occurs and on the speed of the acceleration mechanism. On the other hand, protons of the same energy do not suffer significant energy losses and can freely penetrate into interplanetary space. There are indications that protons with the highest energy can produce the white-

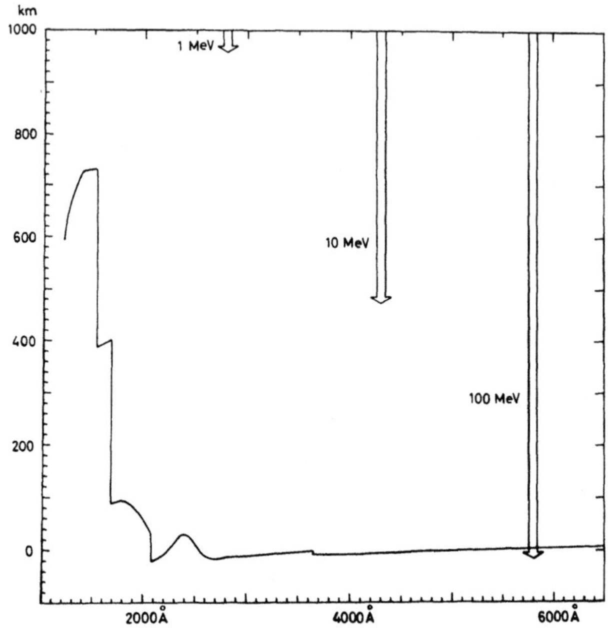


Fig. 6. Height in the solar atmosphere at which $\tau=1$ for different wavelengths (zero level corresponds to $\tau=1$ at 5000 Å) and the depth to which protons of different energy penetrate. (After Švestka, 1970b.)

light emission and ultraviolet continuum bursts, which are observed simultaneously with the hard X-ray burst and radio microwave maximum (Švestka, 1970b).

Figure 6 shows the height in the solar atmosphere, in which – at different wavelengths – the optical depth is unity (Gingerich and de Jager, 1968), and the depths to which protons of different energy penetrate (Schatzman, 1965). If a significant part of the proton energy is transformed into heating of the colliding gas, a short-lived increase in the continuum brightness should occur as actually has been observed in several events, and it should be most pronounced in the ultraviolet. If this is true, then the continuum flash is a direct evidence that protons are accelerated to sub-relativistic energies at the time of the hard X-ray burst and, moreover, *the white-light emission shows us the place where the acceleration and the hard X-ray burst occur.*

## 2.4. The Magnetic Field Configuration

Let us put a question: What is the reason that some flares are associated with an impulsive acceleration process and some others are not? There, most probably, the decisive factor is the magnetic field configuration in the flare region.

An illuminating example has been presented by Švestka and Simon (1969, Figures 3a, b, c), who have studied the flare activity in the region that produced two proton flares, on August 28 and September 2, 1966. Before August 26 the flares – and there were many of them – had almost no reaction in the radio waves, hence no significant acceleration processes occurred in the active region at that time. The situation, however, completely changed on August 26, when strong active radio-emission appeared, more than one half of all flares and subflares were accompanied with radio bursts, and the low-energy (>0.6 MeV) proton flux began to increase. Thus, on August 26 acceleration processes began to occur very often in the active region.

What happened in the active region on that particular day? Figure 7 shows the sunspot group in the morning hours of August 26, when southern and northern polarities were distributed quite regularly in the active region. During this day, however, new big spots of northern polarity developed very close to the southern polarity area (cf. the situation on August 27 in Figure 7), the active region became magnetically complex and gradients of the magnetic field in this part of the active region strongly increased.

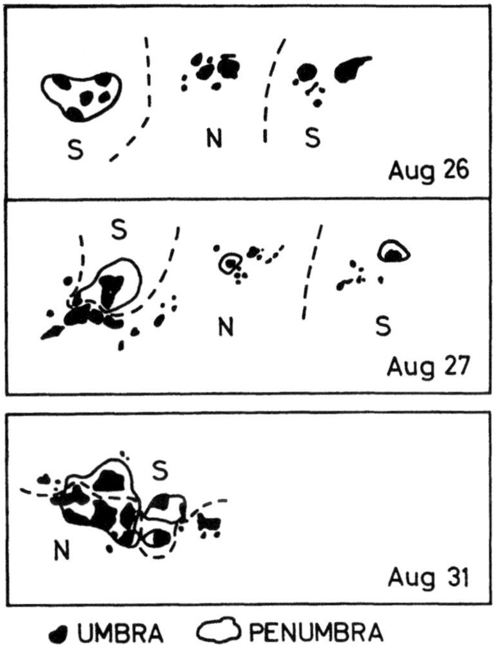


Fig. 7. The sunspot group which produced the proton flare of August 28, 1966, one and two days before the proton flare occurrence (after Mogilevsky *et al.*, 1969). The dashed lines separate the northern (N) and southern (S) magnetic polarities. Note the increased magnetic complexity of the group, and the close approach of spots of opposite polarity, on August 27. The last picture shows the group three days later, when a $\delta$-configuration developed (spots of opposite polarity within a common penumbra). Another proton flare occurred in the region on September 2.

The magnetic complexity of the group and an increase in the magnetic-field gradient (Figure 8), manifested by a close approach of spots of opposite polarity, are the most general characteristics of active regions in which proton flares are produced (Severny, 1963, 1968; Gopasyuk *et al.*, 1963; Avignon *et al.*, 1963 and 1965). Another characteristic feature is an embedding of spots of opposite polarity within common penumbra, the so-called $\delta$-configuration (Warwick, 1966), which also developed in this active region a few days later (cf. the situation on August 31 in Figure 7). Such conditions are always related to at least two solar centers evolving very close one to the other or penetrating one into the other (Martres, 1968). Therefore, they can be realized most easily in long-lived complexes of solar activity, where the probability of such a coincidence in positions is higher than in other parts of the solar surface, as actually has been observed (Švestka, 1968a). This fact may be of advantage for long-term forecasting of the proton flare occurrences as soon as we better understand the rules which govern the large-scale magnetic-field distribution on the sun.

### 2.5. Development of the Proton-Active Region

Generally, the characteristic conditions described above begin to develop two or three days before the proton flare occurrence. These days of preparation (Figure 8) are also characterized by an intensification and a hardening both of the microwave

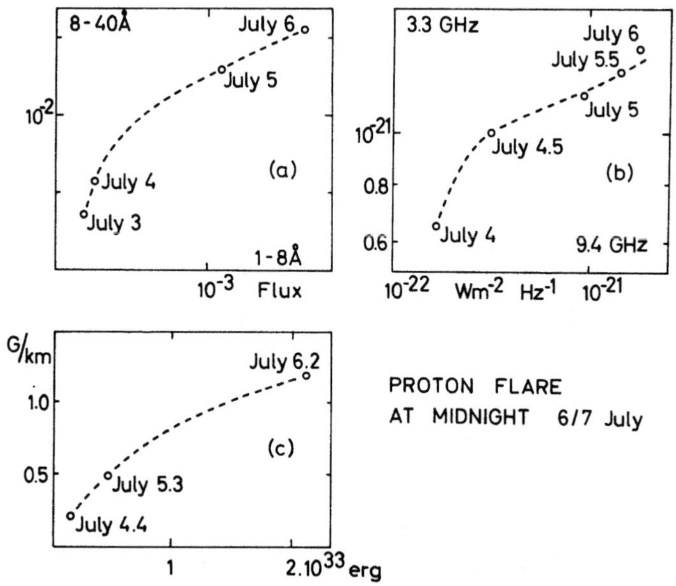


Fig. 8. (a) The X-ray flux increase and its hardening preceding the cosmic-ray flare of July 7, 1966 (after Michard and Ribes, 1968). (b) The radio flux increase in the microwave range, stronger at shorter wavelengths, during the same period (after Tanaka *et al.*, 1969). (c) The increase in the magnetic energy (ergs) and in the magnetic field gradient (G/km) during the same period (after Severny, 1968).

(Tanaka and Kakinuma, 1964; Tanaka *et al.*, 1969), and X-ray spectrum (Friedman and Kreplin, 1969; Křivský and Nestorov, 1968; Michard and Ribes, 1968; Banin *et al.*, 1969; Švestka and Simon, 1969). In the microwave range the usual maximum close to 10 cm is suppressed by the very much increased flux at shorter wavelengths. The increase in X-ray flux often exceeds one order of magnitude in the spectral range of a few angströms, decreases with the increasing wavelength and becomes insignificant for wavelengths above 40 Å. During the same time, the association of flares with X-ray and microwave bursts is strongly increased (Eliseev and Moiseev,

1965; Eliseeva, 1967; Křivský and Nestorov, 1968), which indicates that small acceleration processes already occur in the active region, while the area is preparing for a production of a big particle event several tens of hours later. All these facts suggest that *some particular magnetic configuration must be formed, before significant acceleration processes can occur.*

### 2.6. A TENTATIVE PICTURE OF THE ACCELERATION PHASE

When a proton flare develops in the region, it usually has a typical shape of two roughly parallel bright ribbons, which first form along the line $H_{\|}=0$, which separates the polarities, and then move away one from the other, first fast and then with a decreasing velocity, until they meet and cover large spots. The phase of the fast expansion coincides in time with the acceleration process as established before from the radio and X-ray data (Křivský, 1963; Valníček, 1967; Zirin and Russo Lackner, 1969; Švestka and Simon, 1969; Křivský and Švestka, 1970).

We can tentatively visualize the whole process in the following way (Švestka, 1970b): Figure 9 shows drawings of the sunspot group which produced the proton flare of May 23, 1967. Spots of opposite polarity have been embedded within a common penumbra, fairly close one to the other, and the main part of the proton flare formed along the zero-line of the magnetic field between these spots. All the drawings refer to the same time, coinciding with the maximum of the microwave radio burst and hard X-ray burst, hence to the phase when the acceleration process was in action. The two bright ribbons observed in the Hα line (Figure 9b) can be considered for visible roots of a system of loops, which formed a hot thermal condensation above the sunspot group (Figure 9c). As the condensation was expanding, an impulsive acceleration process set in in that part of the condensation where the magnetic-field gradient was the highest (Figure 9d). Characteristics of the volume in which the acceleration took place, are given in the upper right corner of the figure: In a region of coronal condensation, where density can be estimated to the order of $10^{10}$ particles per $cm^3$ and temperature to several tens of million degrees, one in fifty particles is accelerated to energy exceeding 100 keV. But the size of this volume, where the acceleration occurs, is only a small fraction of the total size of the condensation formed above the flare region. The accelerated electrons produced an impulsive X-ray burst through bremsstrahlung, and a microwave burst by synchrotron radiation in the strong magnetic field, in which the acceleration occurred. Accelerated protons penetrating to the lowest chromosphere and upper photosphere gave rise to a short-lived white-light emission at the roots of the loop system, inside of which the acceleration occurred (Figure 9e).

The size of the volume in which the acceleration is accomplished determines the quantity $N$, and depends on the magnetic-field configuration. In the vast majority of cases it is much smaller than in the event described, and only small particle events are recorded. In other, rather exceptional cases, the volume is very extensive, or several such volumes form inside the hot condensation, not necessarily all at the same time, and then we meet with a particle event of great intensity. Of course, apart

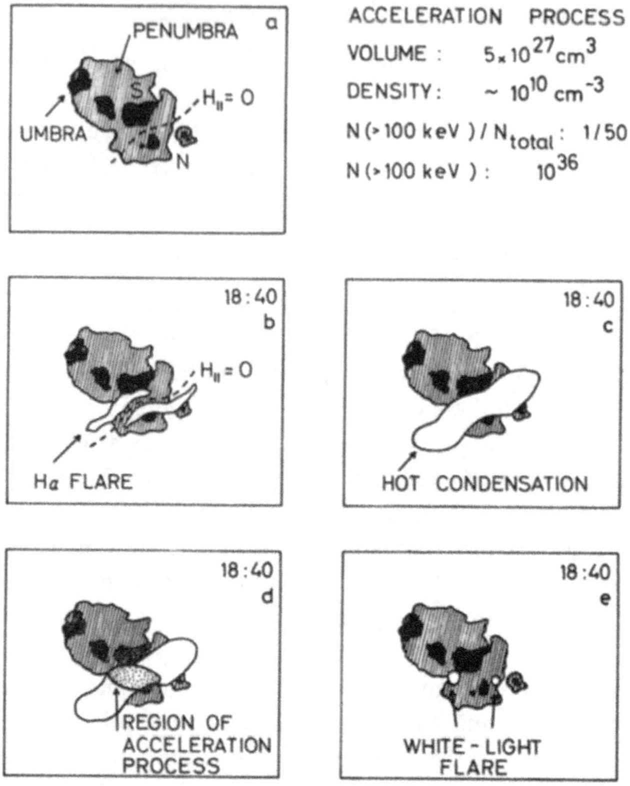


Fig. 9. Drawings of the sunspot group which produced the proton flare of May 23, 1967. (a) The $\delta$-configuration. (b) The brightest parts of the two bright ribbons in the H$\alpha$ line going parallel to the $H_{\parallel} = 0$ line and partly covering the sunspots. (c) The core of the expanding hot condensation composed of a system of loops, the feet of which produce the bright ribbons in the H$\alpha$ line. (d) The region of the impulsive acceleration process in the part of the condensation, where the magnetic-field gradient was the highest. (e) White-light flare emission at position of the roots of loops inside of which the acceleration process occurred. (a), (b), (e) observed (De Mastus and Stover, 1967); (c) and (d) hypothetical.

from it, also the efficiency of the acceleration process, which determines the parameter $\gamma$, certainly is different from one case to the other, depending again on the detailed magnetic configuration in the region where the acceleration occurs. The resulting strength of the particle event as recorded at the earth or aboard a spacecraft, then depends on the position of the particle source on the sun relative to the detector, and – apart from it – on local propagation conditions in the corona above the active region and in interplanetary space.

## References


Anderson, K. A.: 1969, *Solar Phys.* **6**, 111.

Arnoldy, R. L., Kane, S. R., and Winckler, J. R.: 1968, *Astrophys. J.* **151**, 711.

Avignon, Y., Martres, M. J., and Pick, M.: 1963, *Comptes Rendus* **256**, 2112.

Avignon, Y., Caroubalos, C., Martres, M. J., and Pick, M.: 1965, in R. Lüst (ed.), 'Stellar and Solar Magnetic Fields', *IAU Symp.* **22**, 373.

Baker, M. B., Santina, R. E., and Masley, A. J.: 1969, 'Modeling of Solar Cosmic Ray Events Based on Recent Observations', AIAA Paper No. 69–14.
Banin, V., de Feiter, L. D., Fokker, A. D., Martres, M. J., and Pick, M.: 1969, *Ann. IQSY* **3**, 229.
Castelli, J. P. and Aarons, J.: 1969, 'Radio Burst Spectra and the Short Term Prediction of Solar Proton Flares', Paper presented at XV Electromagnetic Wave Propagation Meeting, Quebec, Canada, September 1969.
Castelli, J. P., Aarons, J., and Michael, G. A.: 1967, *J. Geophys. Res.* **72**, 5491.
Cline, T. L. and McDonald, F. B.: 1968, *Solar Phys.* **5**, 507.
Croom, T. L.: 1970, '19 GHz Solar Radio Bursts as Indicators of Proton Events', paper presented at the Int. Symp. on Solar-Terrestrial Physics, Leningrad 1970.
Datlowe, D. W., L'Heroux, J., and Meyer, P.: 1969, Paper presented at the 11th Int. Conf. on Cosmic Rays, Budapest.
De Mastus, H. L. and Stover, R. R.: 1967, *Publ. Astron. Soc. Pacific* **79**, 615.
Dodson, H. W. and Hedeman, E. R.: 1969, *Solar Phys.* **9**, 278.
Dodson, H. W., Hedeman, E. R., Kahler, S. W., and Lin, R. P.: 1969, *Solar Phys.* **6**, 294.
Eliseev, G. F. and Moiseev, I. G.: 1965, *Izv. Krymsk. Astrofiz. Observ.* **34**, 3.
Eliseeva, L. A.: 1967, *Izv. Krymsk. Astrofiz. Observ.* **37**, 153.
Fan, C. Y., Pick, M., Pyle, R., Simpson, J. A., and Smith, D. R.: 1968, *J. Geophys. Res.* **73**, 1555.
Friedman, H. and Kreplin, R. W.: 1969, *Ann. IQSY* **3**, 78.
Fritzová-Švestková, L. and Švestka, Z.: 1966, *Bull. Astron. Inst. Czech.* **17**, 249.
Gingerich, O. and De Jager, C.: 1968, *Solar Phys.* **3**, 5.
Gopasyuk, S., Ogir, M., Severny, A. B., and Shaposhnikova, E.: 1963, *Izv. Krymsk. Astrofiz. Observ.* **29**, 15.
Hultqvist, B.: 1969, in C. de Jager and Z. Švestka (eds.), *Proceedings Symposium on Solar Flares and Space Research*, North-Holland, Amsterdam, p. 215.
Kahler, S. W. and Lin, R. P.: 1969, *Ann. IQSY* **3**, 299.
Kane, S. R.: 1969, *Astrophys. J.* **157**, L139.
Kinsey, J. H. and McDonald, F. B.: 1968, in K. O. Kiepenheuer (ed.), 'Structure and Development of Solar Active Regions', *IAU Symp.* **35**, 536.
Koechlin, Y., Raviart, A., Tregner, L., Blaud, C. J., Degli Antoni, G., Dilworth, C., Maccagni, D., and Tanzi, E. G.: 1969, Paper presented at the 11th Int. Conf. on Cosmic Rays, Budapest.
Křivský, L.: 1963, *Bull. Astron. Inst. Czech.* **14**, 77.
Křivský, L. and Nestorov, G.: 1968, *Bull. Astron. Inst. Czech.* **19**, 197.
Křivský, L. and Švestka, Z.: 1970, 'The Proton Flares of June and November 1968 and February 1969', *Space Res.* **10**, 817.
Lin, R. P.: 1970, *Solar Phys.* **12**, 266.
Lin, R. P. and Anderson, K. A.: 1967, *Solar Phys.* **1**, 446.
Lindgren, S. T.: 1968, *Solar Phys.* **5**, 382.
Malville, J. M. and Smith, S. F.: 1963, *J. Geophys. Res.* **68**, 3181.
Martres, M. J.: 1968, in K. O. Kiepenheuer (ed.), 'Structure and Development of Solar Active Regions', *IAU Symp.* **35**, 25.
McCracken, K. G.: 1969, in C. de Jager and Z. Švestka (eds.), *Proceedings Symposium on Solar Flares and Space Research*, North-Holland, Amsterdam, p. 202.
McCracken, K. G., Rao, U. R., and Bukata, R. P.: 1967, *J. Geophys. Res.* **72**, 4293.
Michard, R. and Ribes, E.: 1968, in K. O. Kiepenheuer (ed.), 'Structure and Development of Solar Active Regions', *IAU Symp.* **35**, 420.
Mogilevsky, E. I., Demkina, L. B., Dolginova, Yu. N., Ioshpa, B. A., Obridko, V. N., Shelting, B. D., and Zhulin, I. A.: 1969, *Ann. IQSY* **3**, 222.
Pick, M.: 1971, this volume, p. 61.
Rao, U. R., Allum, F. R., Bartley, W. C., Palmeira, R. A. R., Harries, J. A., and McCracken, K. G.: 1969, in C. de Jager and Z. Švestka (eds.), *Proceedings Symposium on Solar Flares and Space Research*, North-Holland, Amsterdam, p. 267.
Sakurai, K.: 1969, 'The Generation of Solar Cosmic Rays and the Spectra of Solar Radio Type IV Bursts', GSFC Rep. X-615-69-431.
Schatzman, E.: 1965, in *The Solar Spectrum* (ed. by C. de Jager), D. Reidel, Dordrecht, p. 313.
Severny, A. B.: 1963, *AAS-NASA Symposium on the Physics of Solar Flares*, p. 95.

Severny, A. B.: 1968, in C. de Jager and Z. Švestka (eds.), *Proceedings Symposium on Solar Flares and Space Research*, North-Holland, Amsterdam, p. 38.
Smart, D. F. and Shea, M. A.: 1969, 'Proposed Solar Proton Classification System', Paper presented to the IUCSTP Meeting, London, January 1969.
Švestka, Z.: 1966, *Bull. Astron. Inst. Czech.* **17**, 262.
Švestka, Z.: 1968a, *Solar Phys.* **4**, 18.
Švestka, Z.: 1968b, *Solar Phys.* **4**, 361.
Švestka, Z.: 1969, *Solar Phys.* **8**, 400.
Švestka, Z.: 1970a, 'Solar Particle Events', *Space Res.* **10**, 1.
Švestka, Z.: 1970b, *Solar Phys.* **13**, 471.
Švestka, Z. and Simon, P.: 1969, *Solar Phys.* **10**, 3.
Tanaka, H. and Kakinuma, T.: 1964, *Rep. Ionosph. Space Res. Japan* **18**, 132.
Tanaka, H., Kakinuma, T., and Enome, S.: 1969, *Ann. IQSY* **3**, 63.
Valníček, B.: 1967, *Bull. Astron. Inst. Czech.* **18**, 249.
Warwick, C. S.: 1966, *Astrophys. J.* **145**, 215.
Zirin, H. and Russo Lackner, D.: 1969, *Solar Phys.* **6**, 86.

# SOLAR RADIOBURSTS

A. BOISCHOT
*Observatoire de Meudon, France*

The energy transported by radiowaves is known to be very small as compared to that in the other parts of the electromagnetic spectrum, especially the visible light, UV and X-rays. However the emission of radio bursts accompanying solar activity is a very important phenomenon for the solar physicist. The main reason is that the radiowaves are generally emitted by energetic electrons, which can not be studied by many other means. Moreover, these bursts are often emitted high in the corona, where no other possibility exists up to now to detect solar activity.

If we wish to interpret the radiobursts, for deriving the physical properties of the corona, the number and energies of radiating electrons, etc., the theories of the emission mechanisms must be known. A great deal of work has been done in this field during the last decade and, qualitatively, most of the bursts are now understood. But we are still far from a quantitative interpretation.

## 1. Mechanisms of Emission

We know three mechanisms by which electrons radiate radiowaves in the solar atmosphere: normal bremsstrahlung (or free-free transitions), magnetic bremsstrahlung and conversion of plasma waves.

Radioastronomers express the intensity of a source by its brightness temperature $T_b$, which is the temperature of a black body of the same size which radiates the same energy as the source at the observed frequency. The intensity of the burst at a wavelength $\lambda$ is then proportional to $T_b/\lambda^2$.

For *thermal free-free emission*, it can be shown that

$$T_b = T_e(1 - e^{-\tau}), \tag{1}$$

where $T_e$ = electron temperature, and $\tau$ = optical depth of the source.

If the medium is optically thick ($\tau \gg 1$) as is generally the case for microwaves, we have $T_b \approx T_e$. If the medium is optically thin ($\tau \ll 1$), (1) becomes $T_b \approx \tau T_e$.

In any case, $T_b$ is always smaller than $T_e$. For this reason the normal bremsstrahlung by thermal electrons is not efficient enough to explain the observed high intensities of radiobursts.

But if we consider a non-maxwellian distribution of electrons, i.e. the presence of suprathermal electrons, Equation (1) is still valid if we use for $T_e$ the equivalent temperature of these thermal electrons. Then we can obtain higher values of $T_b$, and of the radiated intensity.

Radiation by *magneto-bremsstrahlung* is that of electrons gyrating around magnetic lines of force in circular or elliptical orbits. It is known as gyromagnetic emission

*Dyer (ed.), Solar Terrestrial Physics / 1970: Part I*, 87–98. *All Rights Reserved.*

if the kinetic energy of the electrons is small, and synchrotron emission if we have to do with relativistic electrons. The two cases lead to quite different characteristics of the radiation.

For gyromagnetic emission the electrons radiate at the Doppler-shifted gyrofrequency and its lower harmonics. The Doppler-shifted gyro frequency is given by

$$f_G = f_H \frac{\sqrt{1-\beta^2}}{(1-\mu\beta\cos\phi\cos\theta)},$$

where $f_H$ = cyclotron (Larmor) frequency, $\beta = v/c$, $\phi$ = pitch angle, $\theta$ = angle between the line of sight and magnetic field, $\mu$ = refractive index.

The higher the energy of the electrons, the higher the orders, $n$, of the harmonics radiated, but in any case, one electron always radiates in discrete emission bands. The computation of the energy and spectrum emitted by a complex source is very difficult, because we have to consider a distribution of electrons with different energies, different pitch angles, an irregular magnetic field, etc, and then have to integrate over $\beta$, $\phi$, $\theta$, $n$ and $H$.

For relativistic electrons (synchrotron radiation) of energy $E$, the orders of the harmonics are so large that the emission is a continuum, around a frequency proportional to the product $H \cdot E^2$. The energy radiated by one electron is proportional to $H^2 \cdot E^2$.

Very large intensities can be obtained if $E$ and $H$ are large enough. It can be shown that, from the observed spectrum of the emission, we can obtain, under certain assumptions, some information about the energy spectrum of the electron distribution. For instance, for a homogeneous magnetic field and a power law distribution of electron energies

$$N(E)\,\mathrm{d}E = KE^{-\gamma},$$

the spectrum of the synchrotron emission follows also a power law

$$I(f) = K'f^{-\alpha},$$

with

$$\alpha = (\gamma - 1)/2.$$

If $T_e'$ is the equivalent temperature of the energetic electron distribution, normal magneto-bremsstrahlung can never lead to a brightness temperature $T_b$ larger than $T_e'$.

But it has been realised recently, that much more energy could be obtained if the electrons radiate coherently, for instance if an organized stream of particles travels through a magneto-active plasma. This phenomenon is referred to by different authors as coherent synchrotron, negative absorption, amplification, or gyromagnetic instability.

The *plasma oscillations* are longitudinal waves which may be excited in a plasma by fast electrons emitting either individually (Čerenkov effect) or collectively (stream instability). They are excited near the plasma critical frequency,

$$f_0 = \sqrt{N_e^2/\pi m} \quad \text{or} \quad f_0 = 9\sqrt{N_e} \quad \text{(MKS)},$$

and cannot propagate through the interplanetary medium. They have first to be converted into electromagnetic transverse waves, by scattering on plasma inhomogeneities.

The theory still in use for this conversion has been proposed by Ginzburg and Zhelezniakov. The plasma waves can be scattered either by neutral inhomogeneities (i.e. density fluctuations) of the plasma, when they give rise to an electromagnetic wave at the same frequency $f_0$; or by space charge irregularities (for instance, thermal fluctuations of plasma waves), to give an electromagnetic wave at the harmonic frequency $2f_0$ of the plasma frequency. Qualitatively this theory seems correct, but there are still many problems to be solved to explain the different characteristics of the bursts. One of them is related to the directivity of the emission. The theory predicts that, in a smooth symmetrical corona, the electromagnetic waves coming from the level near $\mu=0$ can leave the sun only in a radial direction. This is contradicted by the fact that the bursts which are attributed to plasma oscillations (type II and type III) can be observed on the limb as well as near the center of the disk.

## 2. The Radiobursts

With these mechanisms in hand, we shall try to interpret the many different shapes of the radiobursts.

One of the first important results of observation is the big difference between bursts observed in the microwave range (i.e. mm and cm waves) and those at metric and longer wavelengths (Figure 1).

The former are always broad-banded, and vary rather smoothly with time: very rarely we observe emission structures lasting less than a few seconds. On the contrary meter bursts are in many cases short-lived (less than 1 sec) and narrow-banded.

The origin of this difference is due to the difference in density at heights where the bursts are emitted. The microwave bursts come from the lower corona, or upper chromosphere where the density is large enough to absorb all the emissions originating near the critical height (where $\mu=0$). But high in the corona, from where the meter waves originate, the density is small and collisions will have no or little effect on the propagation of radiowaves.

There is probably another reason. The magnetic field in the upper chromosphere and lower corona is large over centers of activity, and many high energy electrons are kept trapped in these regions. They are scattered by inhomogeneities of the magnetic field leading soon to a quasi-isotropic distribution of velocity directions. The emission, averaged over all directions, is then quite smooth and without much fine structure. On the contrary, electrons which can leave these regions, either because their energy is large enough or because they are accelerated along open lines of force, and electrons which are accelerated at greater altitudes, will be bunched into more or less narrow

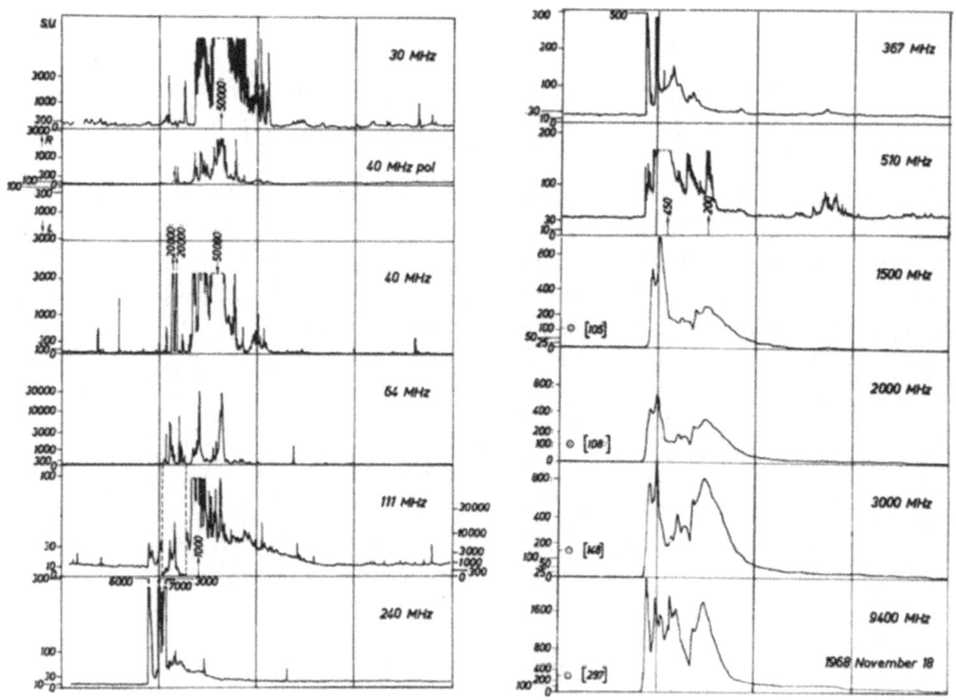


Fig. 1. Emission of radiowaves accompanying a large flare, observed at cm, dm, m and dam wavelengths. The emissions are much more irregular at long wavelengths (Heinrich-Hertz Institute, Berlin).

streams which will interact with the ambient plasma by a number of 'instabilities', giving narrow-band and short-lived emissions.

In the decimetric range, both types of bursts exist and that gives very irregular bursts which have been sometimes taken as distinct types.

### 2.1. Bursts in the Microwave Region

Different kinds of bursts are observed in the microwave range. They can be classified into three main types: gradual bursts, impulsive bursts and type IV bursts (Figure 2).

The *gradual burst* is characterized by a gradual increase to a small peak intensity, and a slow decrease to the original level. It might be observed either isolated (gradual rise and fall) or associated with impulsive bursts, before this burst (precursor) or after it (postburst). The low intensity and other characteristics are easily explained by a thermal bremsstrahlung mechanism: this means that we have to do with transitory heating of small dense regions above the centers of activity: the source of the bursts is of the same nature as the longer lived condensations which give the well-known slowly varying component associated with young centers of activity.

The *impulsive bursts* show a rapid increase and a much slower, quasi-exponential

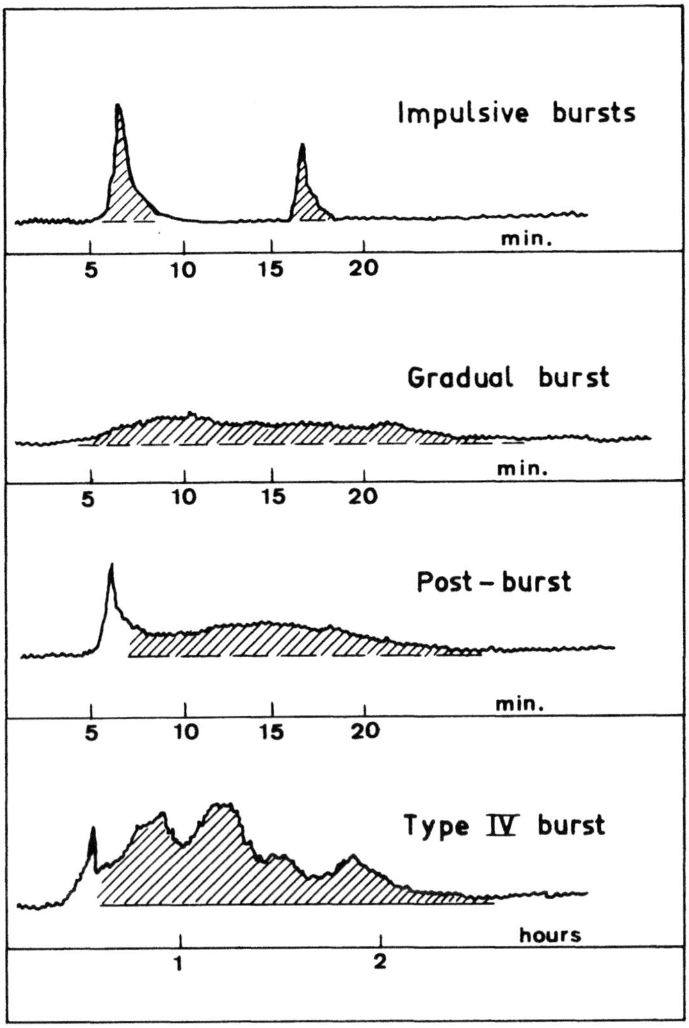


Fig. 2. Some examples of microwaves gradual and impulsive bursts.

decrease, the total duration being between some tens of seconds to several minutes. The high intensities of some of them cannot be explained by a thermal mechanism. They are due either to bremsstrahlung by suprathermal electrons, or to gyromagnetic radiation by mildly relativistic electrons (10 keV to 100 keV).

In both cases, the impulsive bursts are due to energetic electrons. This explains the good correlation which has been found between occurrence of these bursts and that of the X-rays bursts which are also thought to be due to bremsstrahlung by suprathermal electrons.

### 2.2. Type IV burst and relativistic electrons

A type IV burst is a very complex solar emission accompanying large flares only.

It is a continuum, with or without fine structure, which extends from millimeter to decameter wavelengths. Several mechanisms are at the origin of type IV bursts. From the observation of source positions, we can distinguish three main components: the microwave type IV, the metric moving part (type IVA) and the metric fixed part (type IVB).

The first two are due to synchrotron emission by relativistic electrons. But it is difficult to derive the number and energies of the electrons from the observed intensity, because we have only a relation between these quantities and the intensity of the magnetic field.

The interpretation of the observed spectrum of type IV components (Figure 3) is not easy, due to the effect of the coronal plasma upon the emission and propagation of the radio-waves:

(1) All the frequencies do not come from the same source, because only the waves with $f > f_0$ can propagate.

(2) The plasma has an action on the emission itself and we can no longer use the classical formula of synchrotron emission in vacuum. This effect, known as the Razin effect, adds a low frequency cut-off, given, for the relativistic case, by

$$f_R = 20(N_e/H).$$

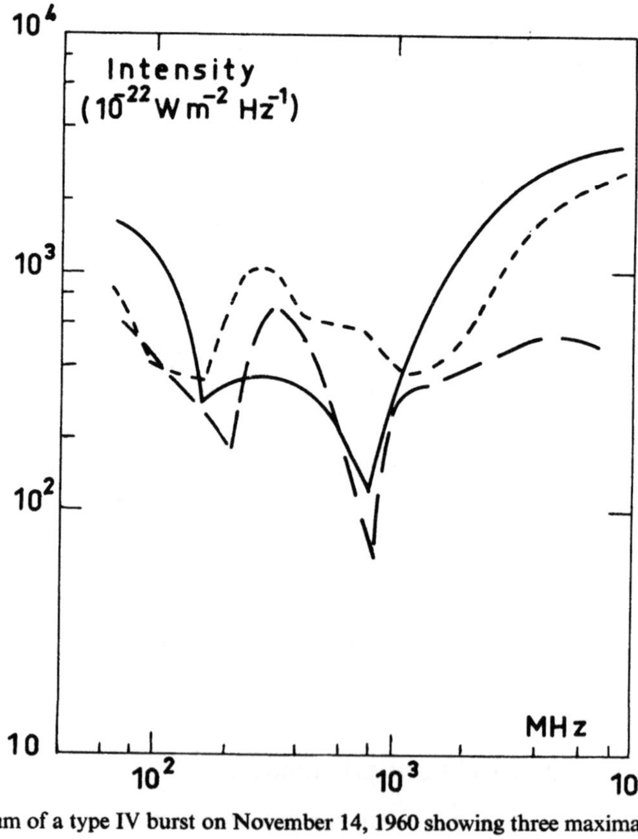


Fig. 3. Spectrum of a type IV burst on November 14, 1960 showing three maxima (after Takakura).

This explains the sharp cut-off toward low frequencies which is often observed in metric type IVA. It was shown by Boischot and Denisse (1964), that a large type IV burst on metric wavelength can be accounted for by $10^{33}$ to $10^{34}$ electrons of 3 MeV in a field of 1 gauss. On the other hand Takakura (1967) explains the microwave type IV by electrons of 100 keV, in a field of 100 gauss. Electrons for microwave type IV bursts are certainly accelerated in the flare region itself, but those which give the metric type IVA might be of a different origin. We have even indications that sometimes these electrons are accelerated at high altitudes.

Type IVB bursts observed at metric wavelengths are certainly of a different nature. They can last for hours, and probably extend down to decametric waves. The polarisation is generally large, corresponding to the ordinary mode, if we refer to the polarity of the magnetic field of the leading spot.

It has been proposed that this component is due to Čerenkov plasma emission, probably like the continuum emission of noise storms. But we are not now in a position to derive the number and energy of electrons involved in this process.

### 2.3 Bursts on Meter and Decameter Wavelengths

These bursts are generally short-lived, covering a narrow band of frequencies and can best be classified from their dynamic spectra. The classification is mainly due to J. P. Wild and the Australian group. They identified five types of bursts, I to V, as shown on Figure 4. The type IV has already been described. The origin of type V is still unknown. It is a short continuum on metric wavelengths following some high intensity type III bursts.

#### 1. *Type III and streams of electrons*

The type III is a burst drifting rapidly from high to low frequency, covering all or part of the range from 500 MHz, to 500 kHz. It is observed either isolated or, more often, in groups of 10 to 100 bursts. Type III bursts are flare associated, but they are observed with flares of all importances; when the association is with large flares they coincide with the impulsive phase.

The early theory proposed by Wild is still generally accepted: a type III burst is due to the conversion of plasma oscillations excited near the plasma frequency by a stream of electrons moving upward in the corona. This theory is supported by the fact that the lower the frequency, the higher the altitude of emission in the corona. But some features are still unexplained. For instance, it is known that an emission originating near the plasma level (where $\mu = 0$) is strongly focussed, with the result that type III bursts would be observed only near the central meridian; this is contradicted by the observation of many type III bursts on the limb.

There are also problems with the emission of the first harmonic, which is observed in about 30% of the bursts. But one of the most important problems about Wild's theory is the stability of the electron stream. Sturrock showed that this stream, exciting plasma oscillations, must be damped out in a fraction of a second, which is much less than the total observed duration of the burst. Smith and Tsytovitch recently

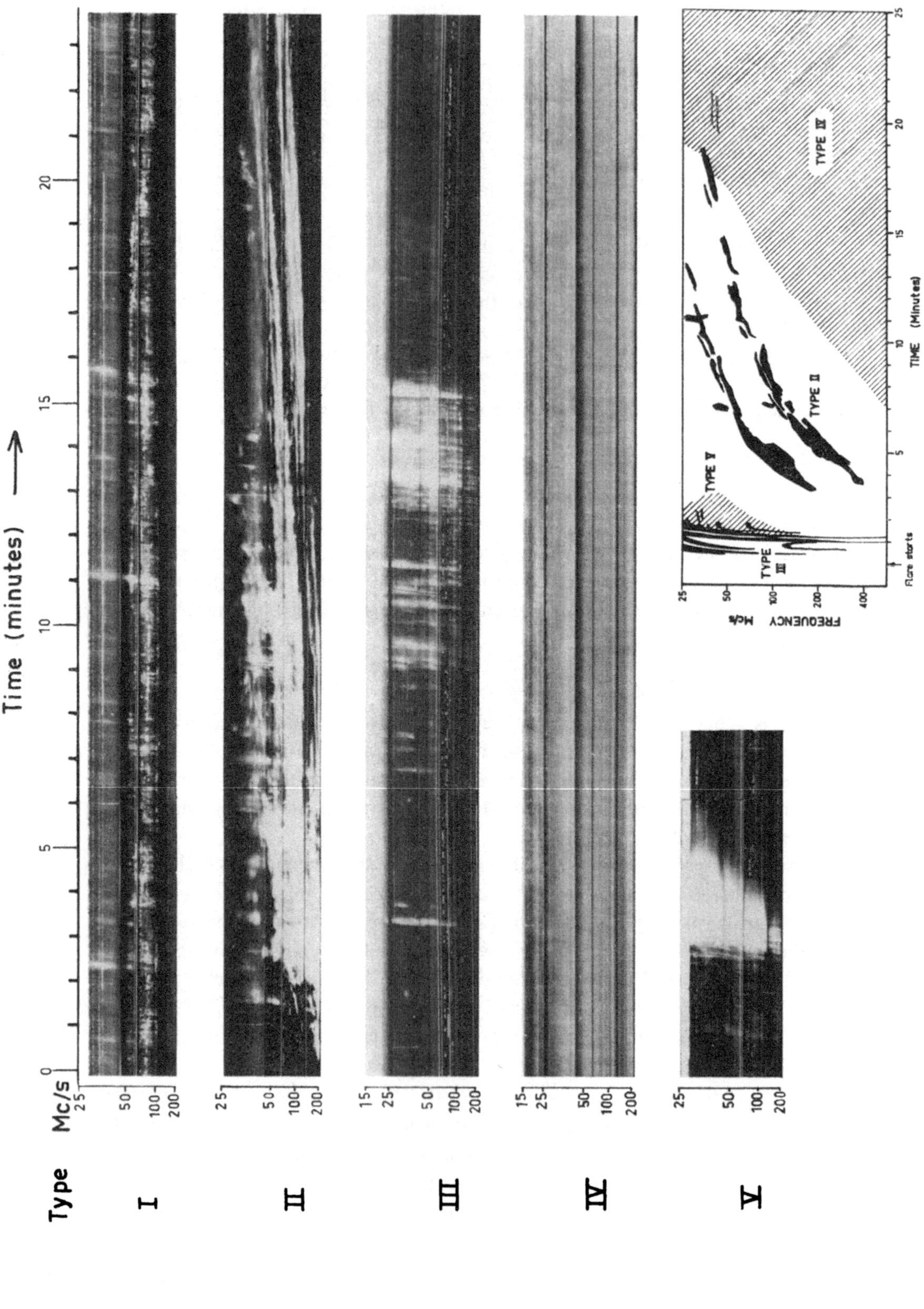
Time (minutes)
Type
Mc/s
I
II
III
IV
V
TYPE III
TYPE V
TYPE II
TYPE IV
FREQUENCY Mc/s
TIME (Minutes)
Flare starts

stressed the importance of non-linear phenomena in the excitation of plasma oscillation, and this may give a clue to the solution of the problem, but to confirm this point a lot of work has still to be done on the non-linear theory.

From the frequency drift rate and a model of the coronal density distribution one can derive that the velocity of the stream lies between 0.2 and 0.5 *c*. This corresponds to electrons of 20 to 200 keV. The total number of electrons, after Wild, would be of the order of $10^{35}$ in each stream, but, in connection with the previous remark about the stability of the stream, this number has to be taken with great care.

Two important conclusions can be drawn from the plasma hypothesis for type III bursts:

The first is about the mechanisms of acceleration of particles in the solar atmosphere. We know very little about these mechanisms, but the sharp rise of type III bursts shows that the mechanism is able to accelerate a very large number of electrons ($10^{35}$?) up to energies of 20 to 200 keV, within a fraction of a second, and that this process can be repeated many times in a period of a few minutes, to give the observed grouping of type III bursts (Figure 5).

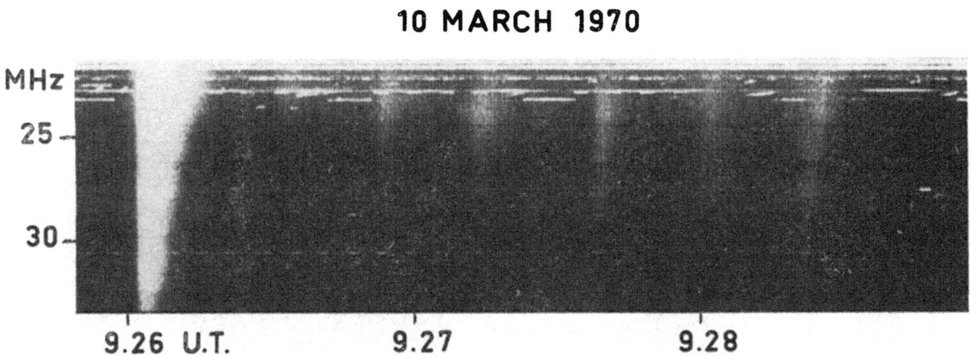


Fig. 5. Periodic occurrence of type III bursts observed from 20 to 40 MHz at Nançay (France).

The second conclusion is that these electrons can leave the sun along open magnetic field lines. Some are stopped in the low corona, below 2 $R_\odot$, giving type III bursts with a sharp cut-off toward low frequencies, some are curved downward and give inverted U bursts, but most of them, which give type III bursts extending to hectometer waves have to leave the sun to reach the corresponding altitudes of more than 20 $R_\odot$. It would be important to get a complete set of observations, with positions of sources, in a large frequency range to demonstrate the relation between these open field lines and the coronal streamers.

## 2. *Type II bursts and shock waves*

Type II bursts are manifestations of quite a different phenomenon. Their dynamic spectrum shows a drift from high to low frequencies, at a rate much less ($\sim$100 times) than type III. The emission has also much fine structure: presence of first harmonic, band-splitting, superposition of very short fast-drifting emissions, etc. The whole phenomenon may last from 1 to 20 minutes.

Wild's theory, assuming that emission occurs near the plasma frequency as for type III, allows one to derive that the velocity of an ascending perturbation lies between 500 and 5000 km/sec. This perturbation can not be a stream of electrons, because this velocity is smaller than the velocity of thermal electrons in a 1 to 2 million degree corona (7000 km/sec). But it is larger than both sonic and Alfvén velocities, and hence, we are very likely in the presence of hydromagnetic shock waves.

This shock is closely related to the wave-like phenomenon sometimes observed running over the surface of the sun and causing 'disparitions brusques' of filaments, triggered flares, nimbus, etc.

However, fast electrons are probably involved in the process which gives the type II emission. The shock wave can not directly excite plasma waves because its velocity is smaller than that of the thermal electrons. Then the theory has been modified by assuming that, in front of the shock wave, electrons are accelerated to much higher energies: it is probably these electrons, leaving the shock upward and downward, which give the type II emission, and the very characteristic herring-bone structure sometimes observed in type II (Figure 6).

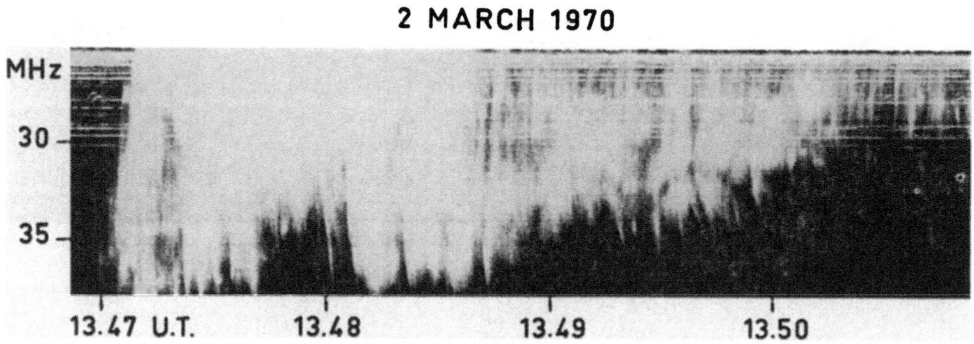


Fig. 6. The herring-bone structure observed in a decameter type II observed at Nançay (France).

However it seems that these electrons do not travel very far from the shock, for the herring-bone structure is relatively narrow-banded. Then there is no possibility to detect these electrons near the earth.

With type II bursts, we meet the same problem as for type III's: the origin of the first harmonic which is observed in 60% of the events. There is moreover a band-splitting of a few MHz of both the fundamental and the harmonic which is still not quite understood.

### 3. *The noise storm and continuous ejection of particles*

The noise storm is the most frequent but the least understood solar emission. It consists of very fast and narrow-banded type I bursts superposed upon an intense continuum at metric wavelengths. We do not yet know the mechanism of this type of bursts: gyro-emission or plasma instabilities; neither where the electrons come from: they can be accelerated either at chromospheric levels or high in the corona where the type I bursts are emitted. Some observations show that the first hypothesis is more likely. If we look at a noise storm on decametric wavelengths, we may see that it consists of a series of type III bursts (Figure 7), even if it lasts for hours or days. The deca-

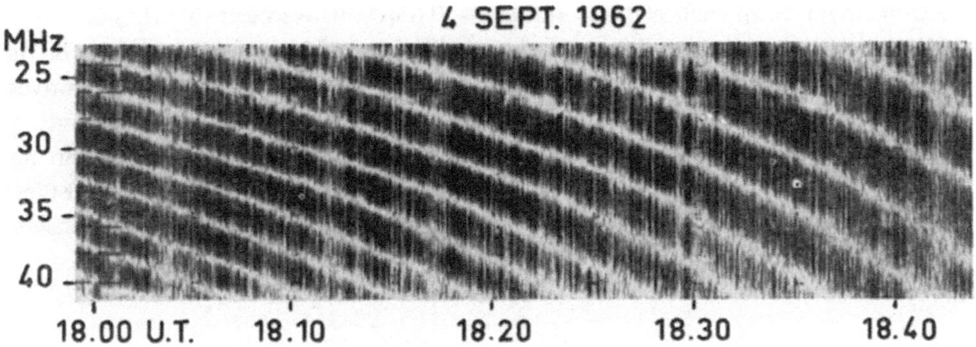


Fig. 7. A noise storm on decametric wavelengths showing a type III structure (the white fringes are instrumental) (after J. W. Warwick).

meter waves are emitted at altitudes of 1 to 2 $R_{\odot}$, and we can be sure that the electrons leave the solar neighbourhood. Then we have to admit that there is a continuous production of fast electrons, or, in other words, that conditions exist in centers of activity that continuously accelerate electrons up to energies of 10 to 100 keV even when there is no flare activity.

## 3. Conclusion

From the point of view of particle production by the sun, three types of bursts are specially important.

The first is the type IV burst which corresponds to a large burst of energetic electrons accompanying flares; these electrons are either trapped in the low corona giving the microwave type IV burst, or escape to higher altitudes to give metric type IV. Protons are accelerated at the same time, as shown by the good correlation between the arrival of protons at the earth and the occurrence and intensity of type IV bursts in the cm and dm range.

The second type is the impulsive microwave burst, which is certainly due to high-energy electrons and which radiates from the same region where the X-rays bursts come from.

The third type of burst related to particle ejection is the type III burst. It has been shown that a stream of electrons which give the type III burst can reach large distances from the sun. But a direct relation between a type III burst and electrons observed near the earth's orbit is difficult to find, due to the very large number of type III bursts that are observed.

A more interesting conclusion about type III bursts is derived from their occurrence in long-lasting storms at decameter waves. This is a proof that even when there is no flare, the centers of activity can be sources of a quasi-continuous flux of electrons. These electrons might be observed near the earth.

The important problem we now meet is to compute the number and energy of electrons involved in each of these emissions. To do this we need two things.

(1) A better quantitative theory of the mechanisms of emission, especially for the excitation of plasma oscillations and their conversion into electromagnetic waves.

(2) Much more information about the size, structure and position of sources at different frequencies. The large Culgoora telescope, which gives images of the sun with a 3′ resolution every second, has already given very important results about these sources, but we urgently need a series of such instruments covering the whole observable frequency range.

## References

Boischot, A. and Denisse, J. F.: 1964, *Advan. Electron. Electron Phys.* **20**, 147.
Kundu, M. R.: 1965, *Solar Radioastronomy*, Interscience, New York.
Takakura, T.: 1967, *Solar Phys.* **1**, 304.
Wild, J. P., Smerd, S. F., and Weiss, A. A.: 1963, *Ann. Rev. Astron. Astrophys.* **1**, 291.
Zhelezniakov, V. V.: 1964, *Radio Emission from the Sun and Planets*, Izd. 'Nauka', Moscow.

# EVIDENCE THAT SOLAR X-RAY EMISSION IS OF PURELY THERMAL ORIGIN (ALSO OBSERVATION OF FAR UV FLASH DURING 28 AUGUST 1966 PROTON FLARE)

TALBOT A. CHUBB

*E. O. Hulburt Center for Space Research, Naval Research Laboratory, Washington, D.C. 20390, U.S.A.*

**Abstract.** X-ray emission from the sun, as thus far observed, is fully interpretable as thermal plasma emission from sets of hot plasmas at different temperatures. Solar minimum conditions are characterized by the presence of only a few plage regions, which dominate X-ray emission below 20 Å. Activity events under these conditions can cause a large increase in flux below 20 Å in coincidence with a decrease in emission between 44 and 60 Å, as the silicon and magnesium ions that dominate 44–60 Å emission move to higher stages of ionization. Under flare conditions temperatures rise to the 10–30 million degree regime, with the higher temperature portions of the plasma dominating the shorter wavelength portions of the spectrum. Thus, flare temperatures calculated from the ratio of S XV and S XVI lines are slightly higher than values calculated from the Si XIII and Si XIV lines, and no major discrepancy exists between these temperatures and temperatures calculated from X-ray continuum emission in the 4–6 Å part of the spectrum. The emission from most flares in the 1–6 Å region is from plasmas that are too cool to contain observable numbers of one-electron iron and calcium ions. The short bursts of very high energy X-rays accompanying strong explosive type flares also show a multithermal spectral character. None of the hard (> 20 keV) X-ray spectra show evidence of monoenergetic electron swarms.

A rare burst of far ultraviolet radiation was observed during the flare of 28 August 1966. The burst lasted four minutes and was coincident with emission of hard (> 80 keV) X-rays and with the visible brightening and disappearance of a small subregion of the flare, as seen in Hα.

Despite the variety of forms which encompass the phenomena called solar activity, these phenomena show a remarkable similarity in emission spectra. By now the X-ray and far ultraviolet emission of a great many events have been examined. The common property, which their emission seems to show, is that in each case the emission between 100 eV and ~100 keV has the spectral appearance of thermal emission from sets of plasmas of varying temperature and emission measure. Although many high energy X-ray emissions have been interpreted as being nonthermal, and, although it is abundantly apparent from the presence of solar proton events that nonthermal processes occur during flare activity, nevertheless, such nonthermal character is *not* evident in the X-ray emission spectra of the events.

The multithermal character of solar emission was first suggested by broadband photometer observations of total X-ray emission from the solar disk. In the days when solar X-ray emission was first crudely explored by use of photometers sensitive to selected bands of X-rays, it was found that the gross features of solar emission could be approximated by gray body fluxes, with higher gray body 'temperatures' being required in the shorter wavelength portions of the spectrum. It was found, for example, that at 44–60 Å a gray body fit to observed intensity data could be obtained by using a $5 \times 10^5$ degree gray body curve, whereas a fit to a $2 \times 10^6$ degree curve was needed for the 2–8 and 8–20 Å regions. Attempts to use higher temperature curves in the 44–60 Å region resulted in excess fluxes in the 8–20 Å region. Attempts

*Dyer (ed.), Solar Terrestrial Physics / 1970: Part I, 99–118. All Rights Reserved.*

to use lower temperature curves in the 2–20 Å region failed to give flux distributions extending to short enough wavelengths to match the atmospheric absorption characteristics observed for the incident flux. Furthermore, the hotter emission components were found to be strongly affected by the solar cycle and by solar activity events.

The thermal character of the plasmas responsible for solar X-ray emission became more clearly established with the recording of solar X-ray lines in 1963 (Tousey *et al.*, 1965; Austin *et al.*, 1966). Widing (1966), following the methods used by Pottasch (1964) at longer wavelength, calculated the intensities of line emissions in the 33–70 Å region for optically thin, isothermal, solar plasmas of varying electron temperature. He showed that the calculated spectra satisfactorily matched the observed solar emission spectrum, which was assumed to be a composite of plage and quiet background emission components. The results of his analysis on the effect of electron temperature on emission are shown in Figure 1. These calculations provide a basis for interpreting broad band, 44–60 Å photometer signals in terms of plasma conditions on the sun. The calculations show that a 44–60 Å photometer responds primarily to line radiations from silicon and magnesium ions, namely from Si VIII (61.0 Å), Si IX (55.3 Å), Si X (50.6 Å), and Mg X (57.9 Å). In addition at higher temperatures emissions from Fe XV (52.9 Å) and Fe XVI (50.4 Å) are indicated. The calculations do not include the contributions from Si XI (41.2 Å) and Si XII (44.1 Å), which are strong in the observed spectrum but which were not considered in Widing's initial study, since they involve difficult-to-calculate nondipole collisional excitation.

Our purpose in presenting Widing's results is to show their relation to some plage emission studies carried out from OSO-2. OSO-2 contained photometers which monitored solar emission in bands 1–8 Å, 8–16 Å, and 44–60 Å. The 44–60 Å band includes the lines shown in Figure 1. As Widing's calculations show, the intensities of the lines above 44 Å vary with temperature in a manner such that above two million degrees the dominant line emissions from Si VIII, IX, and X decrease with temperature. At still higher temperatures, emission from all the Si, Mg, and Fe ions contributing to the 44–60 Å flux can be expected to decrease. The reason for this is the following: as the temperature increases, Si, Mg, and Fe shift to increasingly high states of ionization. As a result, at sufficiently high temperature, the number of these ions available to serve as emitters in the 44–60 Å region decreases with temperature.

OSO-2 was flown during the minimum of the solar sunspot cycle. Under solar minimum conditions, X-ray emission from the entire solar disk can be dominated by a single plage. In this case, heating of the X-ray emitting plasma associated with the plage results in a decrease in the 44–60 Å flux and an increase in emission at shorter wavelengths. Several events showing decreases in 44–60 Å emission simultaneous with increases in 1–8 and 8–20 Å emissions were observed by Kreplin and Gregory (1966) from OSO-2. An example is shown in Figure 2. This event is typical of the group. The predictability of this type of behavior on the basis of thermal analysis illustrates how a potentially puzzling phenomenon is explained by the hypothesis that plage emission arises in a truly thermal plasma.

X-ray radiation emitted during large flare-type events has been found to have the

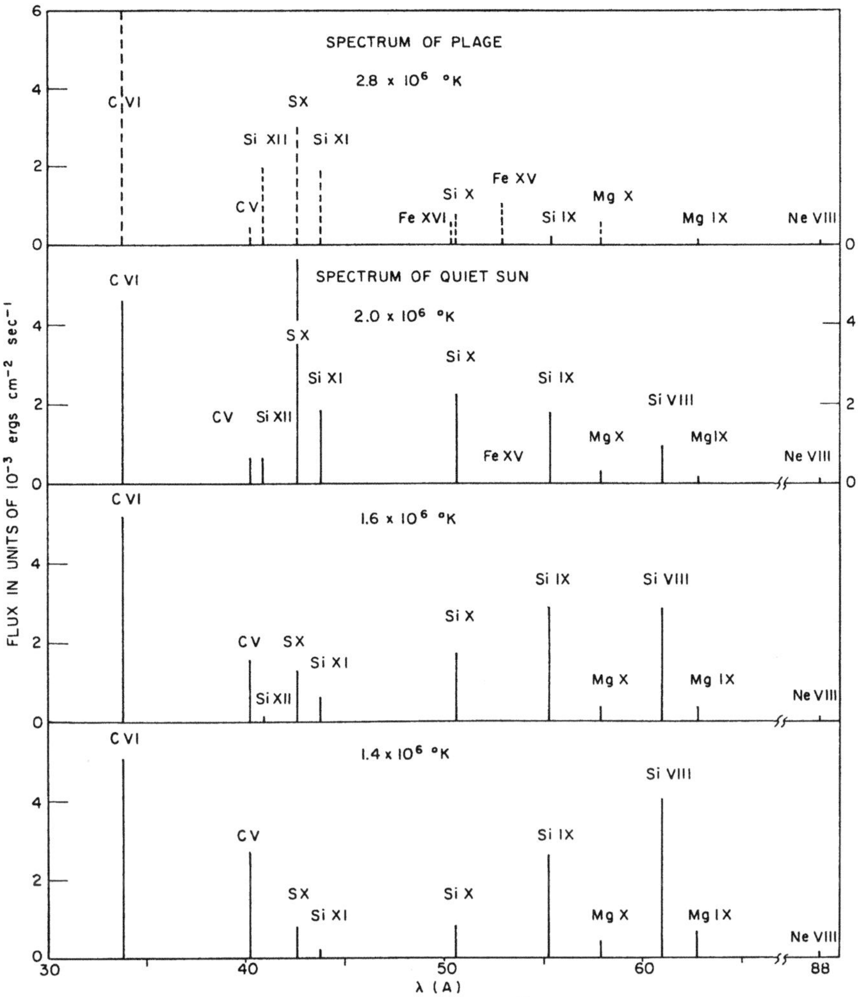


Fig. 1. Theoretical spectra of optically thin solar plasmas as calculated by Widing (1966). The emission measure used for the bottom three spectra is about $2 \times 10^{49}$ cm$^{-3}$, assuming Goldberg *et al.* (1960) abundances. The top spectrum corresponds to an emission measure of about $4 \times 10^{49}$ cm$^{-3}$ on the same assumption. These spectra neglect emission lines from Si XI (49.22 Å) and Si XII (44.02, 44.16 Å). The spectra show how line emissions in the 44–60 Å region can decrease with increasing temperature at temperatures sufficiently high to shift the majority of Si ions to ionization states higher than that of Si XII.

same multithermal character as the integrated X-ray emission from the disk of the sun. Fortunately, the intensity of large flare events is high enough so that useful spectra can be obtained with small Bragg crystal spectrometers. The spectral character of flare X-ray emissions is now well known. The spectra and analyses, presented in Figures 3–7, are largely the result of work by Kreplin, Meekins, and Doschek. In Figure 3, we see the emission spectra of two flares. The spectra have not been corrected for instrumental response. The spectrum of the flare shown by the heavier line is from a large, medium-hot flare observed 16 November 1967; the noisy spectrum shown by the lighter line on the bottom is for a cooler flare observed 24 October 1967. The spectral range shown in the figure is from 1.5 to 8.5 Å, and the data have been plotted logarithmically. The mean temperature for the medium-hot event is about 15 million degrees; the cool event shows a temperature of about 10 million degrees.

Solar flare X-ray emission below 8 Å is dominantly an X-ray continuum emission on which is superimposed a simple line spectrum. Response of the spectrometer to the X-ray continuum shows up discontinuities which are a result of discontinuous changes in the sensitivity of the Bragg spectrometer system. Apparent in the data are discontinuities in sensitivity at the K edge of Al (7.95 Å), at the K edge of silicon (6.74 Å), and at the K edge of Ar (3.87 Å). These changes in level are all instrumental

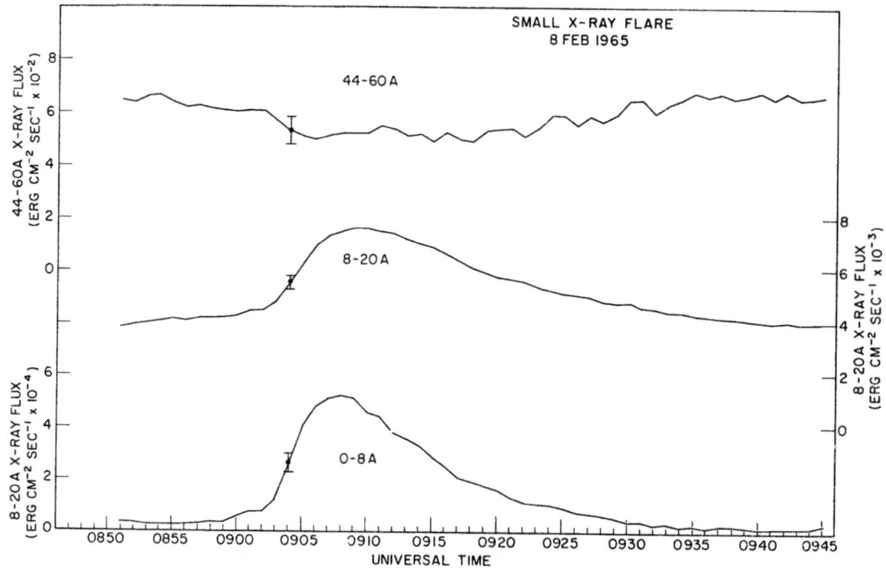


Fig. 2. Time history of X-ray emission of a small X-ray event as observed in three spectral bands from OSO-2. During this period the sun was very quiet, and solar X-ray emissions were probably dominated by emission from a single plage. A decrease in 44–60 Å emission concurrent with an increase in X-ray emission at 8–20 Å and 1–8 Å is considered to be a result of a transient increase in the temperature of the dominant plage and a consequent shift in the mean ionization state of silicon in the plage to states above Si XII. The data plotted are based on uncorrected detector counts, i.e. no correction for short-wavelength response has been subtracted from the 44–60 Å record.

in origin. They are a result of aluminum and silicon in the detector window and in an ultraviolet blocking filter, and of argon in the sensitive gas volume of the X-ray counter. The existence of edges in the data clearly shows that the continuum response is produced by a true X-ray continuum and *not*, for example, by scattered longer wavelength radiation.

Let us now examine the X-ray line spectrum in more detail. The spectrum in the 1.5–8.5 Å region is dominated by emissions from 1 and 2 electron ions of the more commonly abundant medium mass elements on the sun, i.e., Mg, Al, Si, S, Ar, and Ca. There are also significant emissions from Fe and Ni ions containing more than two electrons. The emission from single electron ions corresponds to Lyman line emission in hydrogen. Noted on the figure are, for example, the Lyman-α ($1s\ ^2S$-$2p\ ^2P$), β, and γ lines of one electron Mg (8.42 Å, 7.11 Å, and 6.74 Å). The Lyman-γ line is

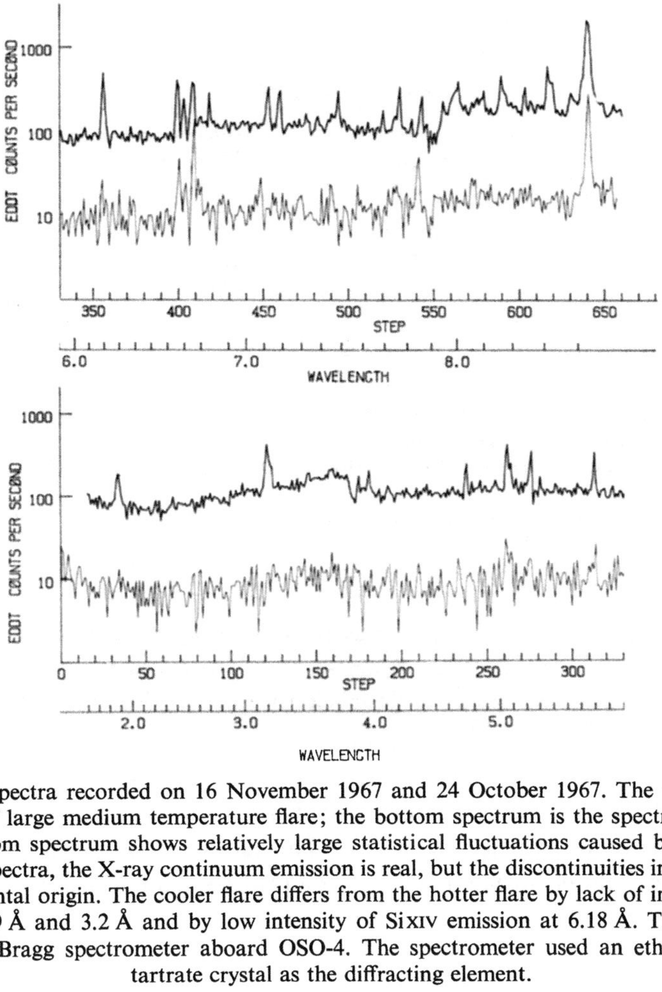


Fig. 3. Flare spectra recorded on 16 November 1967 and 24 October 1967. The top spectrum is the spectrum of large medium temperature flare; the bottom spectrum is the spectrum of a cooler flare. The bottom spectrum shows relatively large statistical fluctuations caused by low counting rates. In both spectra, the X-ray continuum emission is real, but the discontinuities in the continuum are of instrumental origin. The cooler flare differs from the hotter flare by lack of iron and calcium emissions at 1.9 Å and 3.2 Å and by low intensity of Si XIV emission at 6.18 Å. The spectra were recorded by a Bragg spectrometer aboard OSO-4. The spectrometer used an ethylenediamine-*d*-tartrate crystal as the diffracting element.

blended with emission from two-electron silicon. Line emissions corresponding to Al Lyman-α (7.17 Å), Si Lyman-α (6.18 Å), and S Lyman-α (4.73 Å) show up clearly in the spectrum. We do not see Ca Lyman-α (3.02 Å). The lack of Ca Lyman-α is a good indicator that the flare does not contain significant emission measures of plasmas of temperatures greater than 20 million degrees.

Emissions from helium-like ions are also strong. The lines observed include a 'triplet' made up of the resonance, forbidden, and intercombination lines identified by Gabriel and Jordan, plus higher members of the resonance series. The wavelengths of the Gabriel-Jordan (1969) 'triplets' of Ca, Ar, S, Si, Al and Mg are shown in Table I. Also prominent in the spectrum are higher members of the resonance series of two electron magnesium ions at 7.85 Å ($1s^2$-$1s3p$) and 7.47 Å ($1s^2$-$1s4p$) and of two electron silicon ions at 5.68 Å ($1s^2$-$1s3p$).

TABLE I

Strongest components of Gabriel-Jordan 'triplets'

| Element | Transition wavelengths in Å | | |
|---|---|---|---|
| | Allowed [a] ($1s^2\,{}^1S_0$-$1s2p\,{}^1P_1$) | Intercombination [b, c] ($1s^2\,{}^1S_0$-$1s2p\,{}^3P_{1,2}$) | Forbidden [b] ($1s^2\,{}^1S_0$-$1s2s\,{}^3S_1$) |
| Mg | 9.228 | 9.299 | 9.385 |
| Al | 7.802 | 7.853 | 7.924 |
| Si | 6.684 | 6.725 | 6.781 |
| S | 5.062 | 5.089 | 5.126 |
| Ar | 3.965 | 3.984 | 4.010 |
| Ca | 3.187 | 3.201 | 3.219 |
| Fe | 1.855 | 1.861 | 1.869 |

[a] Calculated values; House (1969). Observations by Walker and Rugge (1970) indicate that these values may be high.

[b] Differences between ${}^1P_0$ level and ${}^3P_1$ and ${}^3S_1$ levels, based on Gabriel and Jordan (1969).

[c] This line is actually a close doublet, since the $J=2$ and $J=1$ levels of the ${}^3P$ state contribute to the ${}^1S$-${}^3P$ transition.

The resonance lines of one- and two-electron ions can be used as thermometers to measure the temperature of the emitting plasma. Evans and Pounds (1968) used this approach to study active regions. Doschek (1970), and Walker and Rugge (1970) have studied the resonant $1s$-$2p$ transitions of hydrogen-like and helium-like ions of Mg, Si, S, and Ca. Doschek defines the ratio $R_{\text{Si}}$, for example, to be the intensity of the Lyman-$\alpha$ line of one-electron silicon divided by the line intensity of the resonance $1s^2$-$1s2p$ transition of two-electron silicon. This latter line is the shortest wavelength member of the Gabriel-Jordan 'triplet'. For an isothermal plasma $R_{\text{Si}}$ is related to the plasma temperature by

$$R_{\text{Si}} = \beta \frac{f_{\text{Si}}^{\text{H}}(T)}{f_{\text{Si}}^{\text{He}}(T)} e^{-\Delta E/kT},$$

where $\beta$ is a constant depending on known atomic parameters, $\Delta E$ is the energy difference between hydrogenic and helium-like lines, $f_{\text{Si}}^{\text{H}}(T)$ is the fraction of silicon in the one-electron ion state, and $f_{\text{Si}}^{\text{He}}(T)$ is the fraction of silicon in the two-electron ion state. Since $\Delta E/kT$ is not large, the values of $f_{\text{Si}}^{\text{H}}(T)$ and $f_{\text{Si}}^{\text{He}}(T)$ largely control the value of $R$. The values of $f_{\text{Si}}$ are obtained from ionization equilibrium theory.

Temperatures calculated in the above fashion again indicate the multitemperature character of solar activity plasmas with the hotter plasmas dominating the spectrum at shorter wavelengths. For example, the top spectrum shown in the figure indicates a temperature of 14 million degrees using the silicon line ratio and a temperature of 19 million degrees using the sulfur line ratio.

Doschek and Meekins (1970) have used the resonance line ratio approach to provide a semi-quantitative means of classifying flares. Cool flares are those with line ratios corresponding to temperatures less than 10 million degrees; medium flares are those with temperatures between 10 and 20 million degrees; and hot flares are those with temperatures between 20 and 30 million degrees. The magnesium, silicon, sulfur, and calcium ratios for these classes of flares are shown in Table II. For hot flares the calcium Lyman-$\alpha$ line is fairly prominent, and the calcium ratio is in the range 0.07–0.3. Hot flares appear to be relatively rare.

TABLE II

Flare classification based on temperature

| | Flare description | | | |
|---|---|---|---|---|
| Parameter | Cool | Medium | Hot | Superhot |
| Temperature (K) | $< 10^7$ | $10^7$–$2 \times 10^7$ | $2 \times 10^7$–$3 \times 10^7$ | $> 3 \times 10^7$ |
| $R_{Mg}$ | $< 1.2$ | 1.2–20 | $> 20$ | – |
| $R_{Si}$ | $< 0.2$ | 0.2–3.4 | 3.4–10 | – |
| $R_S$ | $< 0.02$ | 0.02–0.37 | 0.37–1.0 | – |
| $R_{Ca}$ | – | – | 0.07–0.3 | $> 0.3$ |

Figure 4 shows the 1.5–8.5 Å spectrum of the hottest flare we have recorded. The plot is made using a linear scale, which makes the lines appear to stand out more clearly above the continuum. Again we see a large discontinuity of instrumental origin at the argon K absorption edge. This flare is immediately distinguished as a hot flare by the presence of calcium Lyman-$\alpha$ radiation. The spectrum also shows a strong S Lyman-$\alpha$ line, which is comparable in intensity to the resonance line of the sulfur 'triplet'. The silicon Lyman-$\alpha$ line is much more intense than the resonance line of the two-electron silicon 'triplet'.

Figure 5 shows the shorter wavelength portion of the spectrum in greater detail. Again one sees the calcium Lyman-$\alpha$ line and the helium-like calcium 'triplet', which is here clearly resolved. One important feature to note in the figure is the difference in shape between the iron line radiation at 1.9 Å and the calcium radiation at 3.2 Å. The asymmetric shape of the iron emission is due to the presence of strong inner shell emission even in this hot flare. It is estimated that about half of the hot flare iron emission is from KX-ray line emission from iron with three and more electrons, and about half is Gabriel-Jordan 'triplet' emission from two-electron iron. There is no observable emission from one-electron iron ($1s$-$2p$ at 1.79 Å), which is consistent with the 25–30 million degree plasma temperatures indicated by other lines. Similar

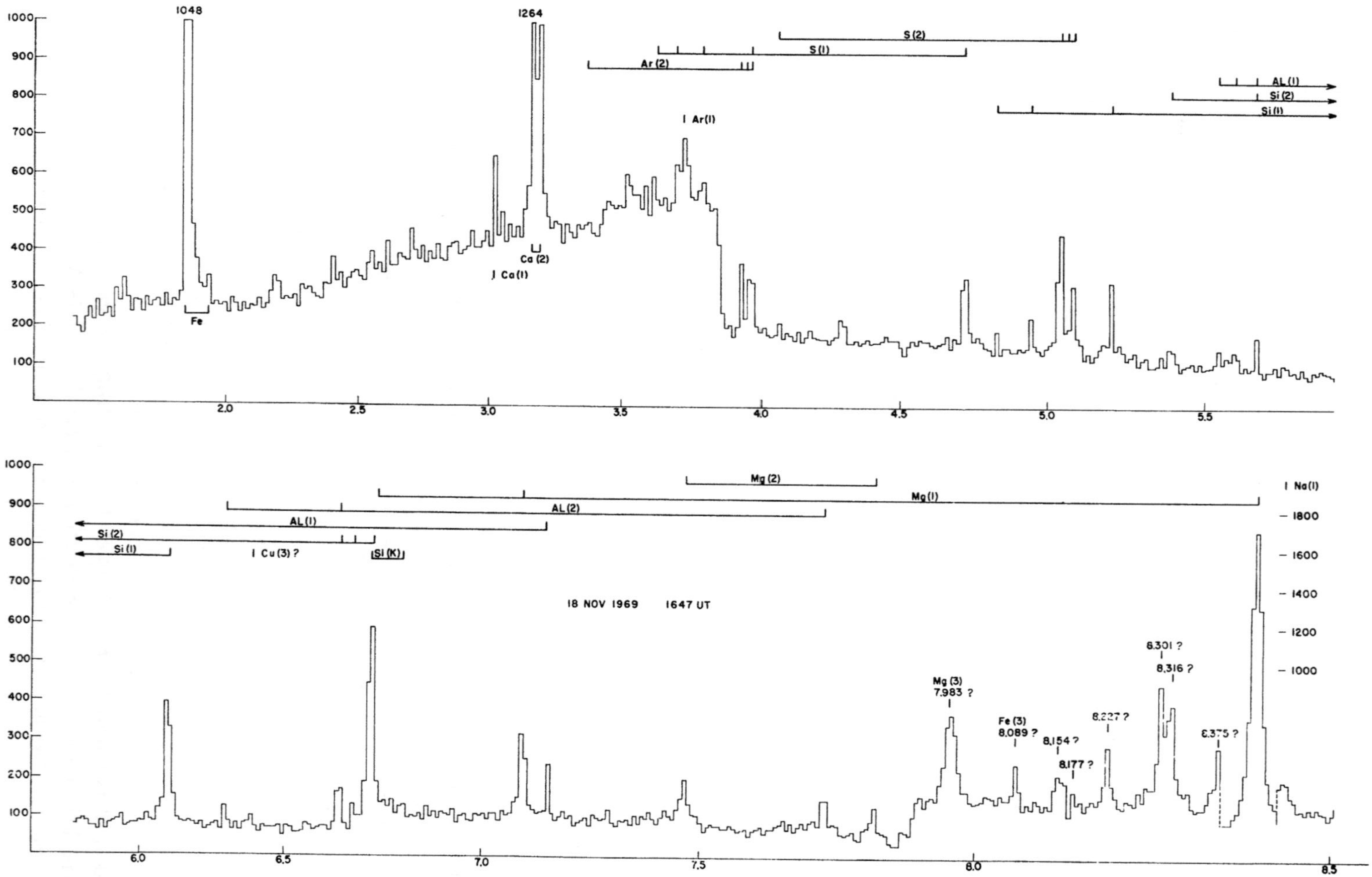


Fig. 4. Spectrum of a hot flare. The high temperature is indicated by such features as the well developed $1s\,{}^1S$-$2p\,{}^2P$ line of Ca XX at 3.02 Å and by the approximate equality of intensity for the $1s\,{}^1S$-$2p\,{}^2P$ line of S XVI at 4.73 Å and the $1s^2\,{}^1S$-$1s2p\,{}^1P^0$ line of S XV at 5.06 Å. The spectrum was recorded by a Bragg spectrometer aboard OSO-6. It is the highest temperature line spectrum thus far observed by OSO-4 or OSO-6 instrumentation.

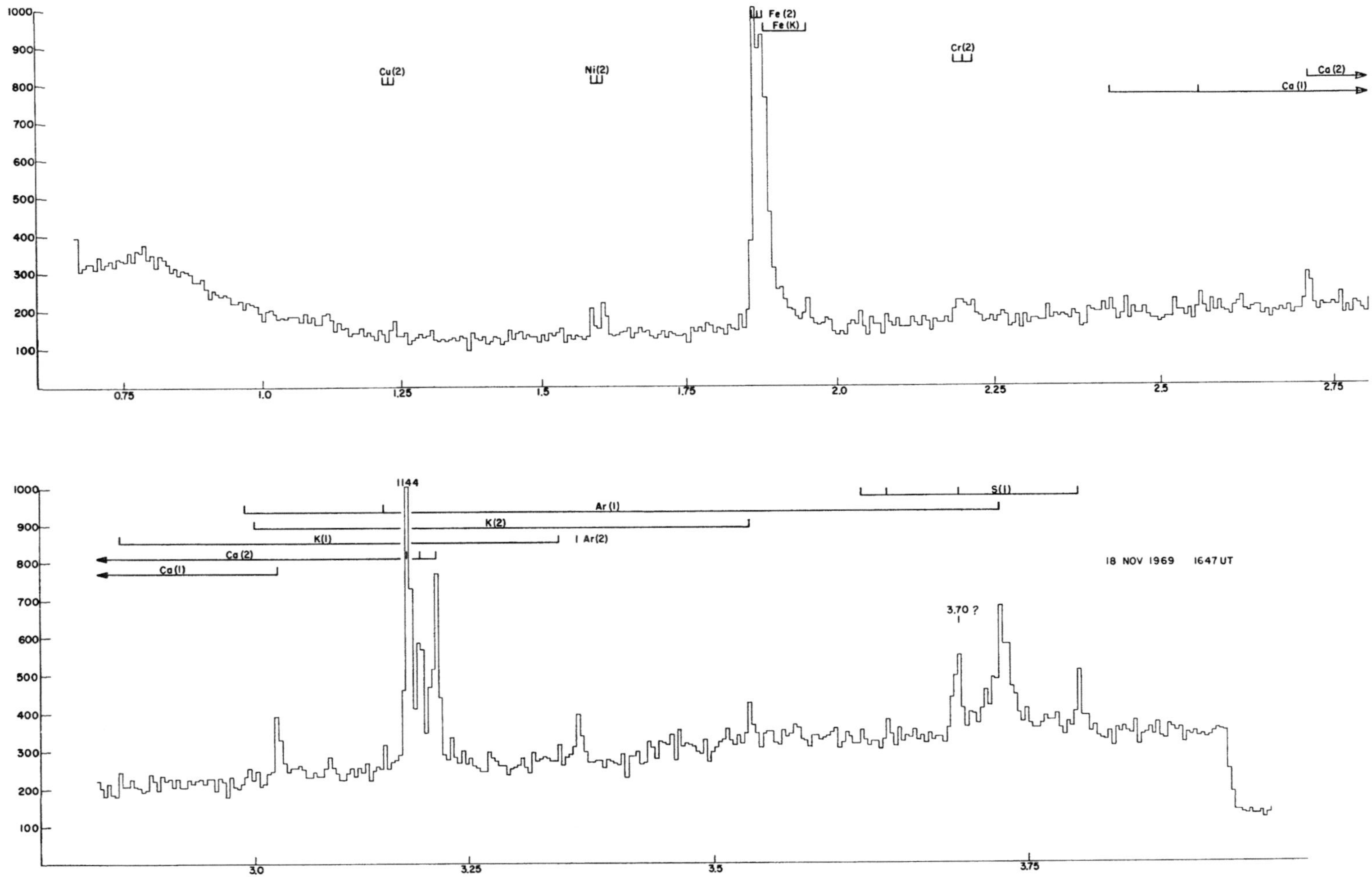


Fig. 5. Short wavelength portion of hot flare spectrum as recorded by a Bragg spectrometer using a LiF diffracting crystal. This spectrum resolves the three components of Ca XIX and indicates the presence of a satellite line as well. The difference in the profile of emission from Fe ions at 1.85 Å and Ca ions at 3.2 Å is a result of the importance of inner shell transitions from Fe XXIV, Fe XXIII, etc. in the iron emission as compared with the dominance of the $1s^2$-$1s2p$ emission from Ca XIX in the calcium emission.

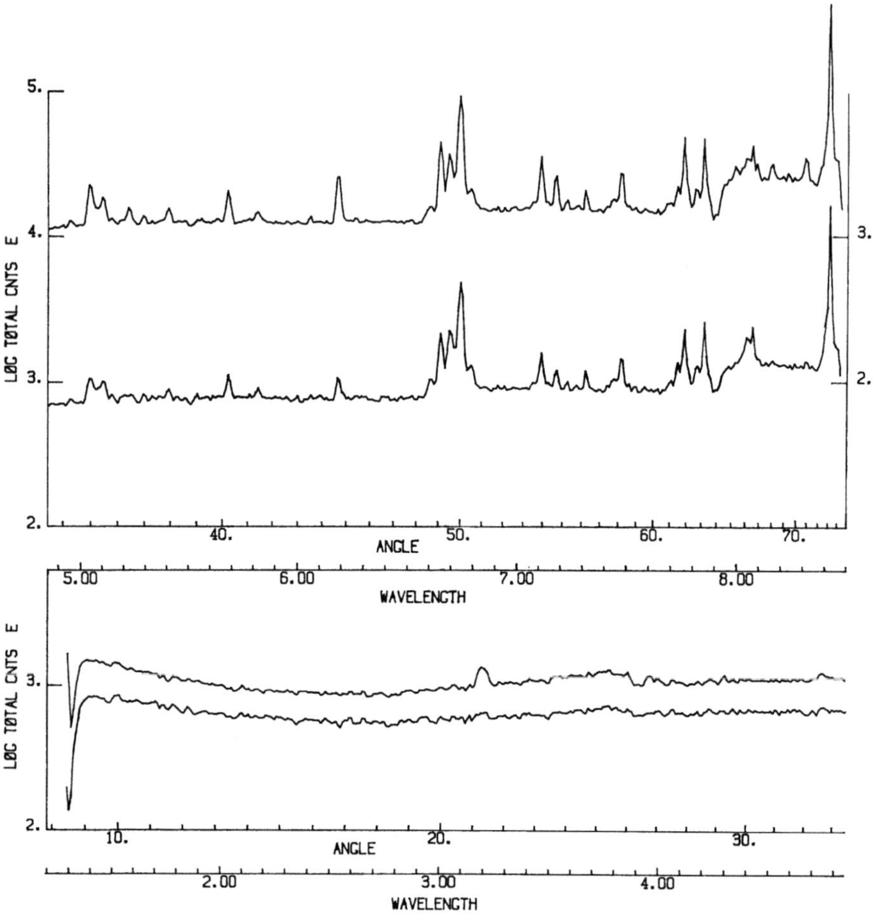


Fig. 6. Composite line spectrum of 90 small X-ray events. The bottom curve shows the summed spectrum of 90 small cool events; the top curve shows the same spectral sum to which has been added the emission of several medium temperature flares. In both spectra the continuum is mostly a summation of particle counts and not a true X-ray continuum. The bottom curve is indicative of a cool flare plasma of $\sim 8.6 \times 10^6$ degrees and shows a Si XIV $1s\,^1S$-$2p\,^2P$ line much weaker than the $1s^2\,^1S$-$1s2p\,^1P^0$ line of Si XIII. The cool flare sum also lacks emission from Ca ions at 3.2 Å. Addition of spectra from a few medium intensity flares to the data set leads to observable Ca emission and increases the relative amount of Si XIV emission.

hot flare spectra have been studied by Neupert and Swartz (1970), and similar large amounts of iron satellite line emission have been recorded.

The bottom curve of Figure 6 shows the line spectrum produced by cool flares. For the cool flares thus far observed, intensities have been sufficiently low that spectra from single events as recorded with our small instruments are quite noisy. Meekins and Doschek, therefore, have summed the emission from 90 low intensity events to

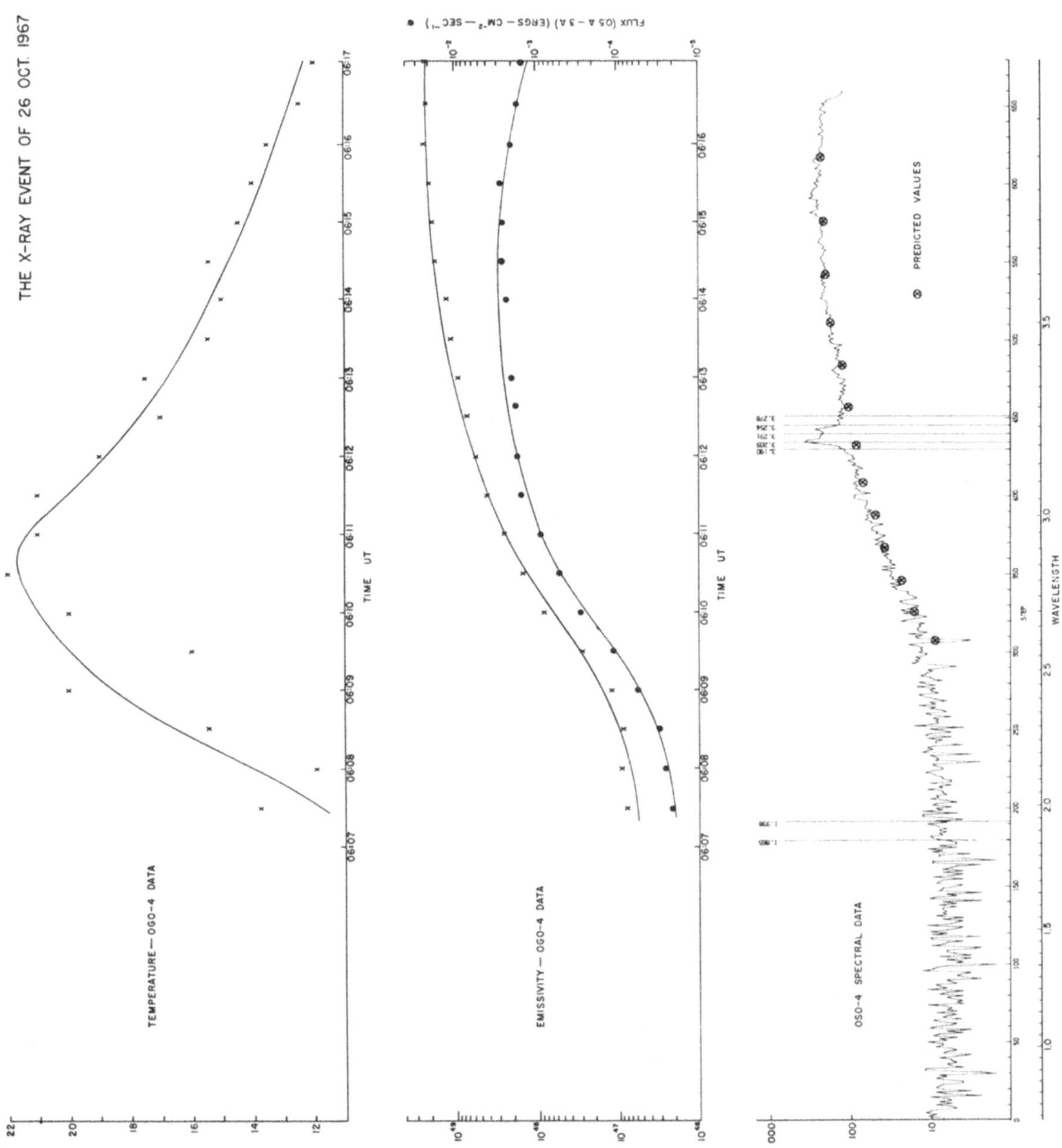


Fig. 7. Time history of flare plasma temperature, emission flux, and emission measure. The X-ray emission history of the flare, as shown in the top three curves was deduced from broad band X-ray measurements in 0.5–3 Å and 2–8 Å bands. This flare shows typical flare plasma behavior: the temperature, energy flux, and emission measure peak in chronological order. Simultaneously recorded spectral data were obtained and compared with the response to be expected on the basis of the broad band photometer readings. The calculations were normalized to the spectrometer data at 3.5 Å. The agreement between the temporal behavior of the calculations and the observed count rates indicates both that the broad band photometers correctly measure average flare plasma parameters and that, except for a wavelength independent efficiency factor of ~ 0.6, the spectrometer response function is understood.

obtain better statistics on spectral lines. One major purpose was to see whether these small events were actually hot events of small emission measure, as opposed to being small cool events. The events were selected on the basis of showing significant Mg Lyman-α emission. The summing procedure results in a build-up of particle produced count so that, in the summed plot, the background count is no longer a measure of pure X-ray continuum, and its slope cannot be used to provide an independent measure of temperature. (The top curve shows the same 90-event sum, except that spectra from several additional medium temperature flares have been added to the sum.) The silicon temperature for the sum of the weak flares corresponds to ~8.6 million degrees; hence, most of the events included were indeed cool events. The sum spectrum shows no identifiable calcium line emission, little or no S Lyman-α, and the absence of a group of lines between 7.8 and 8.4 Å. The latter lines may be due to transitions in iron present in high stages of ionization.

Another method of measuring average temperatures of flare plasmas is to measure solar X-ray emission in two spectral bands and to compare the emission ratio for these bands with emission ratios calculated for plasmas of different temperatures. Such calculations have been carried out by Culhane (1969) and by Horan (1970). The OGO-4 and SOLRAD IX satellites provide records of solar X-ray intensity in spectral bands 0.5–3 Å and 1–8 Å. From these intensities, Horan has obtained time histories of 1–8 Å flux, of electron temperature and of emission measure, assuming that the emission is produced by free-free and radiative recombination processes. A typical time history is shown in the upper part of Figure 7. The emission profile shows that the temperature, X-ray energy flux, and emission measure peak in chronological order. This sequence of emission history is very characteristic of X-ray flares. Often the emission measure decays very slowly after peak, suggesting that at times the flare plasma becomes part of the overall coronal gas.

Doschek has compared flare plasma temperatures determined from broad band ion chamber measurements with temperatures deduced from X-ray continuum slope measurements as recorded by a Bragg spectrometer. The method of comparison is the following: the time history of temperature and emission measure, as deduced from broad band measurements and shown in the upper part of Figure 7, was used to compute the X-ray flux at earth for each wavelength being recorded by a Bragg spectrometer. From this flux and from the spectrometer spectral response function (crystal reflectivity times counter efficiency times ultraviolet blocking filter transmission), the expected spectrometer counting rate was calculated. The expected count rate vs. time was then normalized to the observed count rate at 3.5 Å to correct for relative errors in the absolute sensitivity of the two detecting systems. The normalizing factor was 0.62. The calculated points are the dots shown on the bottom curve of Figure 7, and the line shows the recorded count rates. The agreement for this flare is quite good. Also, the intensities of the calcium line emissions are in agreement with published solar abundances. Similar calculations carried out for a longer-lived flare did not provide quite such a good fit, probably indicating the need for a multithermal analysis.

We must now ask ourselves whether the multithermal character of solar emission extends to still shorter wavelengths; that is, to the hard X-ray flash that accompanies bright flares. The next three figures are old ones showing pulse amplitude spectra measured during a Class $2^+$ flare on 1 September 1959. Figure 8 shows a sequence of pulse amplitude spectra displayed on semi-log plots and covering the range 20–120 keV. The spectra were interpreted as indicating the existence of a high temperature thermal plasma which cooled somewhat during the 5-min period of observation.

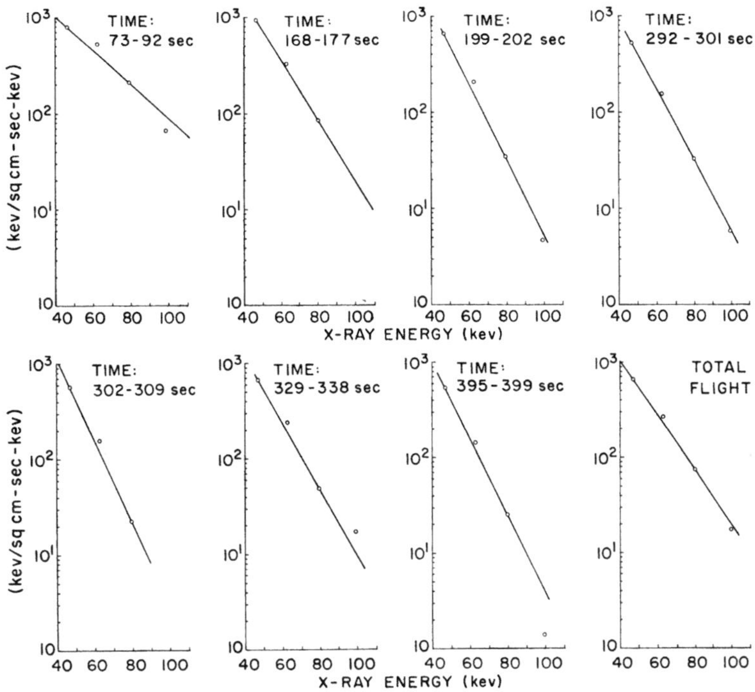


Fig. 8. Pulse amplitude spectra of hard X-ray emitting plasma recorded 1 September 1959. These spectra were interpreted as indicating emission from a cooling, very high temperature plasma. The data were recorded on a scintillation counter spectrometer during a rocket flight.

Figure 9 shows the time history of electron temperature as deduced from the slopes of the spectra. These plasma temperatures are some ten times higher than the temperatures of the plasmas causing line emission between 1 and 8 Å. Figure 10 shows the time history of flare energy flux in the >20 keV photon energy regime and also a history of the hot plasma emission measure. The emission measure of the very hot ($10^8$ K) plasma component is typically some $10^3$–$10^4$ times smaller than the emission measure of the ($10^7$ K) plasma responsible for the 1–8 Å flux of a big flare. On the other hand, if the particle density in the very hot plasma region is sufficiently low, the energy contents of the $10^7$ K and $10^8$ K plasmas could be comparable. Note that the temperature again peaks before the flux, while the emission measure continues to rise, subsequently leveling off at a later time.

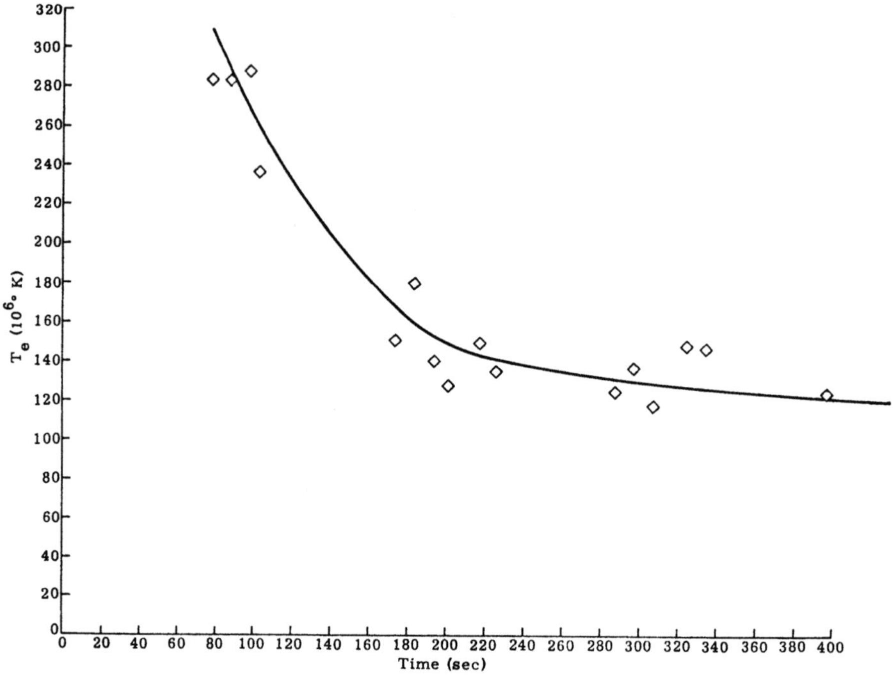


Fig. 9. Temperature history of the very hot plasma component observed in 1 September 1959 flare

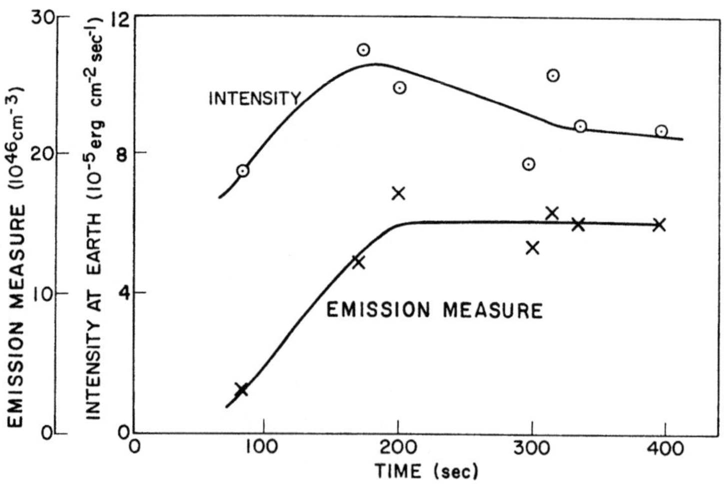


Fig. 10. Time history of hard X-ray flux (> 20 keV) and of emission measure for 1 September 1959 flare. The very hot plasma component of this flare seems to show the same type of behavior as the typical main flare plasma, i.e., first a peaking in temperature, followed by a peaking in flux, followed by a peaking or leveling-off of emission measure.

Much better data on high energy X-rays has been obtained by Frost (1969). Figure 11 displays an example of Frost's data. The figure shows the time history of hard X-ray emission in three energy bands: 14–28 keV, 28–50 keV, and 50–82 keV. On the right is the time history of the same event in three radio spectral bands. The event was a limb event, listed as a Class 3 eruptive prominence with spray. The start of the high energy X-ray event shows a quasi-periodic time variation in X-ray emission, followed by a more regular rise and fall after 2142. Such oscillations were first

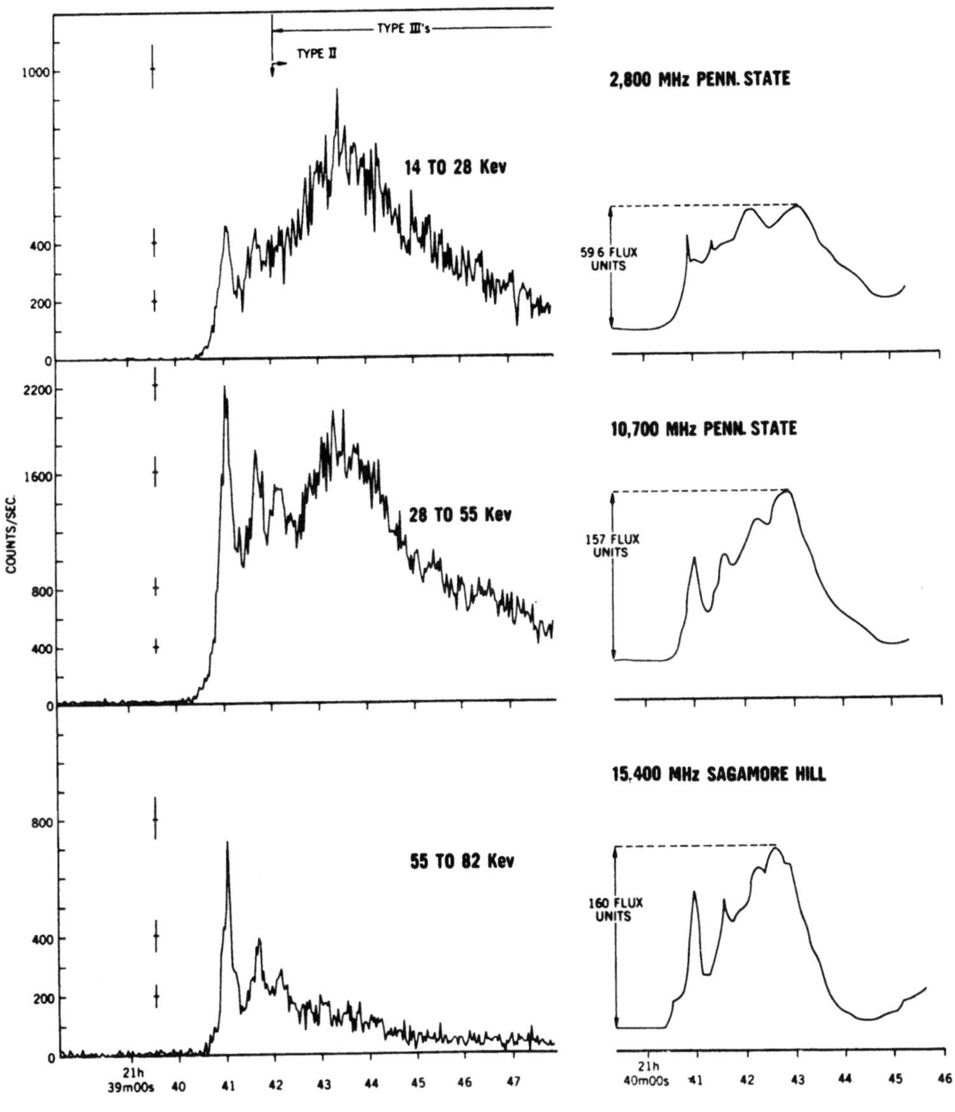


Fig. 11. Time history of hard X-ray emission from limb event of 1 March 1969, as reported by Frost (1969). The X-ray fluxes show an initial oscillatory behavior, followed by a rise and fall. Microwave radio emission histories are shown to the right of the figure and show parallel behaviors.

reported by Parks and Winckler (1969). Figure 12 is a replot of Frost's spectral energy data on a semi-log plot. First we consider the points with filled-in circles. These are the data taken near the peak of the first oscillation at the start of the flare. In his paper on the event, Frost emphasizes the lack of photons of $>300$ keV energy. He interprets the spectrum as a power law spectrum showing a change in spectral index and concludes that a group of electrons was accelerated to roughly a common energy with few, if any, electrons accelerated to energies greater than 300 keV. The X-rays were produced when these electrons interacted with the solar gas. However, the lack of

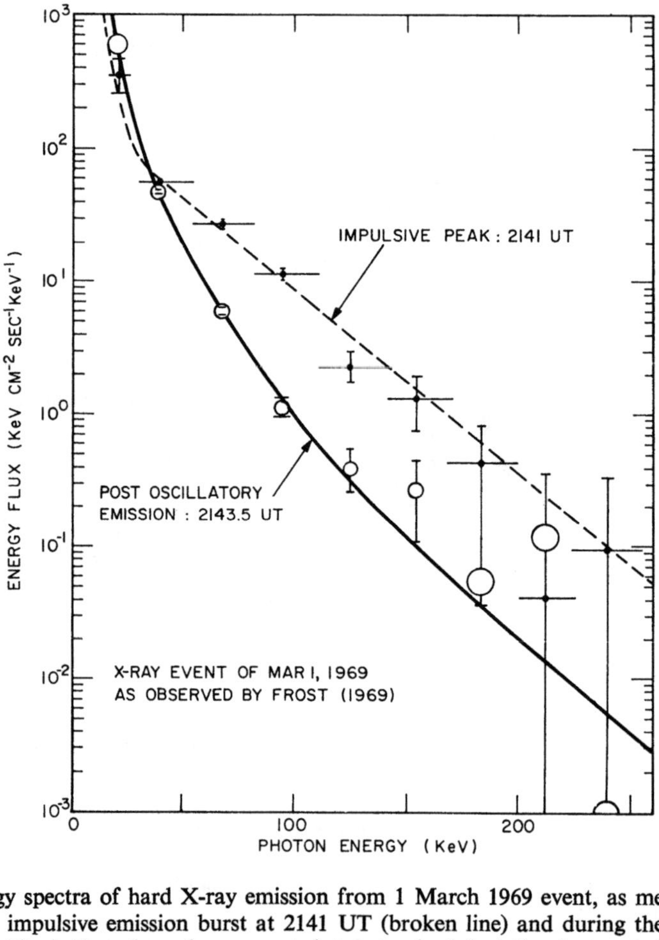


Fig. 12. Energy spectra of hard X-ray emission from 1 March 1969 event, as measured by Frost during the first impulsive emission burst at 2141 UT (broken line) and during the rise and fall at 2143.5 UT (solid line). Error bars show expected $\pm 1$ standard deviation in counts for the 2141 data (dots) and for some of the 2143.5 data (open circles). Where error bars are omitted from the 2143.5 data, large open circles are used to designate the data values, and error spreads similar to the 2141 data apply. The widths of the instrumentation energy channels are indicated by horizontal bars through the 2141 data. Frost interprets the emission as nonthermal, with the 2141 data showing a dual spectral index. The author interprets the 2141 data as free-free emission from a $3.7 \times 10^8$ K plasma with an emission measure of $5 \times 10^{44}$ cm$^{-3}$. The 2143.5 emission is interpretable either as a high temperature multithermal emission or as a nonthermal emission with photon spectral index of 5 (that is, $n(E)\, dE = kE^{-5}\, dE$).

a fall-off in emission at low energy is not suggestive of thick target bremsstrahlung. Instead, a more likely interpretation is that we are again seeing a hot thermal plasma, even during the pulsating part of this X-ray event. Frost's data points fall on a straight line on a semi-log plot with a turn-up at low energy, suggestive of a multithermal plasma situation. A cutoff in emission at high energy is a characteristic of thermal emission. Using the thermal interpretation, we calculate, in accordance with Kahler *et al.* (1970), an emission measure of $5 \times 10^{44}$ cm$^{-3}$ and an electron temperature of $3.7 \times 10^{8}$ K for the hot plasma component at 2141 UT.

Must we then conclude that no nonthermal X-ray emission exists during flare events? This, the author believes may be a premature conclusion. First, the high energy X-rays observed by Frost after the initial bursts could, for example, be interpreted either as a multithermal emission, as shown in Figure 12, or as a nonthermal emission with a power law spectrum, as shown in Frost (1969). Second, detection of low fluxes of very high energy X-rays or of $\gamma$-rays has been reported by Takakura, and fluxes of X-rays of up to 500 keV photon energy have been reported by Cline *et al.* (1968). A thermal explanation of the Cline, Holt, and Hones results is compatible with the observations but requires electron temperature in the $5 \times 10^{9}$ K range.

In view of the spectral measurements that indicate the predominance of thermal X-ray emission, we must ask ourselves, is it reasonable to believe that gas plasmas can be heated to temperatures as high as 400 million degrees without electrons passing through a detectable nonthermal phase? The author thinks the answer is yes. There appear to be many processes that would not produce detectable nonthermal components in the electron energy distribution. One can imagine ohmic heating (although runaways might be expected) or adiabatic compression. On the other hand, it seems possible that in the flare process ions are preferentially energized initially in a process in which magnetic energy is converted into magnetohydrodynamic wave energy, which in turn imparts velocity motion to both ions and electrons. With such a mechanism the ions receive the bulk of the energy, and electrons are heated by subsequent electron-ion collisions. The known large population of nonthermal ions makes this an attractive road to follow.

The remainder of this paper concerns a different subject, namely the proton flare of 28 August 1966. During this event we noted a highly unusual burst of far ultraviolet emission at the start of the flare, as shown in Figure 13. We have not yet seen any other burst of far ultraviolet flux of comparable intensity, although we have not been in position to examine too many flares for this type of behavior. The main burst of ultraviolet emission lasted two minutes, and a second smaller, shorter burst was observed on the decay. The far ultraviolet time history was somewhat similar to that of the 8000 MHz radio noise emission and occurred at the same general time as a burst of $>80$ keV X-rays; however, the initial far ultraviolet burst peaked at a slightly earlier time than either of the other radiations. The far ultraviolet emission was observed by two far ultraviolet photometers, a 1050–1350 Å photometer and a 1225–1350 Å photometer. For both bands the maximum increase in flux at earth was of the order of 0.01 erg cm$^{-2}$ sec$^{-1}$ Å$^{-1}$ (0.013 erg cm$^{-2}$ sec$^{-1}$ Å$^{-1}$ for 1050–1350 Å,

0.0067 erg cm$^{-2}$ sec$^{-1}$ Å$^{-1}$ for 1225–1350 Å, assuming constant energy per Å). The burst occurred in coincidence with a sudden frequency deviation which was reported by Donnelly (1968a, b). This sudden frequency deviation is the result of increased F region ionization, presumably produced by a burst of 100–900 Å radiation. Donnelly estimated the peak intensity of the far ultraviolet flash to be about 7 ergs cm$^{-2}$ sec$^{-1}$ in the 100–900 Å band. Donnelly believes his intensity estimate to be accurate to within a factor of 4. Our ultraviolet intensity estimates may be as good as $\pm 30\%$.

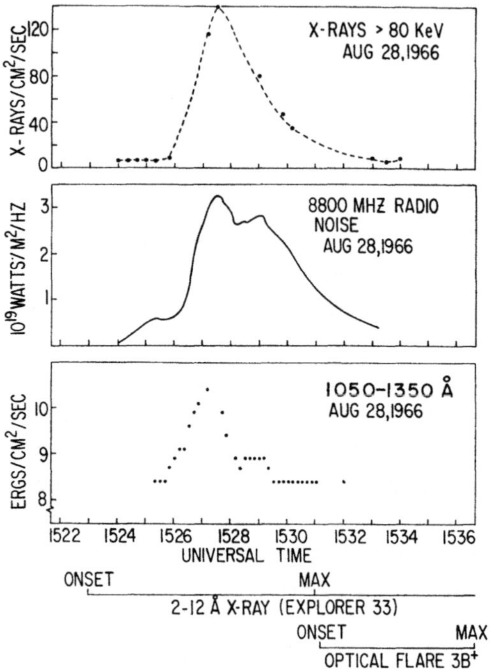


Fig. 13. Time history of short intense emission burst observed during 28 August 1966 flare. The far UV emission burst is the only clearly established burst of such intensity that has been recorded. Comparison between the far UV emission and both the microwave and high energy X-ray emissions indicate a similar time development. The 1050–1350 Å emission was recorded by a photoionization chamber aboard the SOLRAD IX satellite. Although the spectral band includes Lyman-$\alpha$ emission at 1216 Å, the emission increase beginning at 1526 UT is believed not to be due to Lyman-$\alpha$ line emission.

In any case the 1050–1350 Å intensities are based on the same photometry that gave a value of 5.8 erg cm$^{-2}$ sec$^{-1}$ for the post-flare total disk Lyman-$\alpha$ emission of the sun, assuming that 93% of the ion chamber response is caused by the Lyman-$\alpha$ line (as calculated for Hinteregger (1965)).

The far ultraviolet flash occurred at the same time that a small visible subregion of the flare appeared, and both followed the same history. Figure 14 shows the time history of region 5 in the flare, as reported by Dodson and Hedeman (1969). The region was observed at the Hα wavelength and was of very small size, near the resolution limit of optical seeing. It was unusual in being of such short duration and was interpreted by Dodson as being a small bright region of normal flare emission which was quickly obscured by prominence material. However, the coincidence in time history between observations of region 5 and of the far ultraviolet flash suggests that the region may have had an unusual emission character.

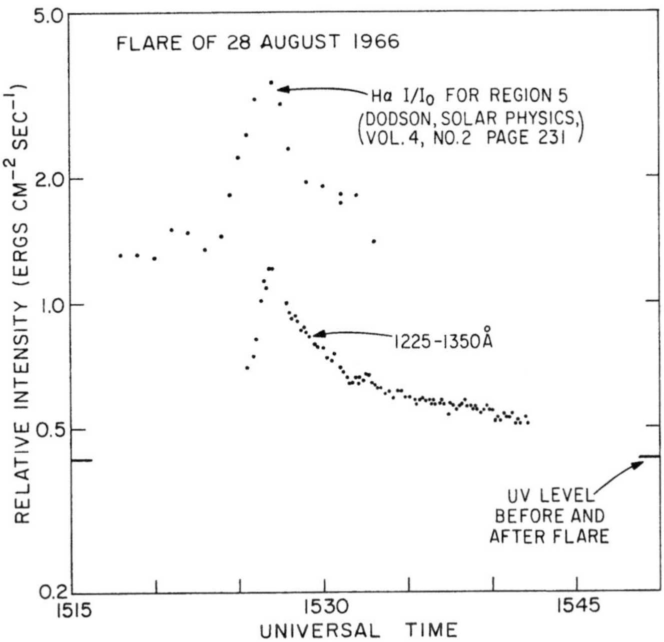


Fig. 14. Time history of short burst of far UV emission observed during 28 August 1966 flare and Hα emission from flare region 5, as reported by Dodson and Hedeman (1968). In this figure, the response of a far UV photometer which excludes the H Lyman-α line at 1216 is shown. The emission of far UV light parallels the visible emission from region 5 observed through an Hα filter. Dodson and Hedeman interpret the rapid fall-off in emission from region 5 as being caused by obscuration from overhead material. An alternate interpretation might be that the visible emission from region 5 and the far UV emission are of common origin.

At this time we have not been able to conclude much concerning the spectral character of the far ultraviolet emission flash. One conclusion is clear; it does not appear possible for the emission to have been broadened Lyman-α line emission, since such emission would have produced a larger current in the 1025–1350 Å ion chamber relative to that produced in the 1225–1350 Å ion chamber than was observed. The ratio is compatible with continuum emission but, also, possibly with emission from a multitude of intense lines. If multiple line radiation is involved, a proportionate increase in the Lyman-α line, however, is again not compatible with the observed detector responses; in other words, higher excitation lines would have to have been involved. The broad spectral range of the emission, however, is suggestive of a continuum which decreases in energy-per-Å as X-ray wavelengths are approached.

The above analysis of the far UV flash from the 28 August 1966 flare is based largely on the work of C. J. Grebenkemper and A. T. McClinton, Jr.

## References

Austin, W. E., Purcell, J. D., Tousey, R., and Widing, K. G.: 1966, *Astrophys. J.* **145**, 373.
Castelli, J. P. and Michael, G. A.: 1967, *Solar Phys.* **1**, 125.
Cline, T. L., Holt, S. S., and Hones, E. W., Jr.: 1968, *J. Geophys. Res.* **73**, 434.
Culhane, J. L.: 1969, *Monthly Notices Roy. Astron. Soc.* **144**, 375.
Dodson, H. W. and Hedeman, E. R.: 1968, *Solar Phys.* **4**, 229.
Donnelly, R. F.: 1968a, *Solar Phys.* **5**, 123.
Donnelly, R. F.: 1968b, ESSA Technical Report ERL 92-SDL 6, Space Disturbances Laboratory, Boulder, Colo.
Doschek, G. A.: 1970, Private Communications on Flare Classification by Line Ratio Temperatures,
Doschek, G. A. and Meekins, J. F.: 1970, *Trans. Am. Geophys. Union* **51**, 415.
Evans, K. and Pounds, K. A.: 1968, *Astrophys. J.* **152**, 319.
Friedman, H.: 1962, *Repts. Progr. Phys.* **25**, 163.
Frost, K. J.: 1969, *Astrophys. J.* **158**, L159.
Gabriel, A. H. and Jordan, C.: 1969, *Nature* **221**, 1969.
Goldberg, L. A., Muller, E. A., and Aller, L. H.: 1960, *Astrophys. J. Suppl.* **5**, No. 45, 1.
Hinteregger, H. E.: 1965, *Space Sci. Rev.* **4**, 461.
Horan, D. M.: 1970, 'Coronal Electron Temperatures Associated with Solar Flares', Ph.D. Thesis, Catholic University of America, Ann Arbor, Mich., University Microfilms, 70-22133.
House, L. L.: 1969, *Astrophys. J. Suppl.* **18**, 21.
Kahler, S. W., Meekins, J. F., Kreplin, R. W., and Bowyer, C. S.: 1970, *Astrophys. J.* **162**, 293.
Kreplin, R. W.: 1961, *Ann. Geophys.* **17**, 151.
Kreplin, R. W. and Gregory, B. N.: 1966, *Space Res.* **6**, 1011.
Meekins, J. F. and Doschek, G. A.: 1970, *Trans. Am. Geophys. Union* **51**, 416.
Meekins, J. F., Doschek, G. A., Friedman, H., Chubb, T. A., and Kreplin, R. W.: 1970, *Solar Phys.* **13**, 198.
Neupert, W. M. and Swartz, M.: 1970, *Astrophys. J. Letters* **160**, L189.
Parks, G. K. and Winckler, J. R.: 1969, *J. Geophys. Res.* **74**, 4003.
Pottasch, S. R.: 1964, *Space Sci. Rev.* **3**, 816.
Tousey, R., Austin, W. E., Purcell, J. D., and Widing, K. G.: 1965, *Ann. Astrophys.* **28**, 755.
Walker, A. B. C., Jr. and Rugge, H. R.: 1970, *Aston. Astrophys.* **5**, 4.
Widing, K. G.: 1966, *Astrophys. J.* **145**, 380.

# PARTICLE ACCELERATION AND PLASMA EJECTION FROM THE SUN

S. I. SYROVATSKY
*P. N. Lebedev Physical Institute, Moscow, U.S.S.R.*

The heating of plasma to high temperatures, the high-velocity plasma ejection, and the acceleration of non-thermal particles are the most outstanding phenomena of solar activity which appear mainly in the course of solar flares. So, in this paper we are interested primarily in solar flares and flarelike phenomena. However, to have a guide to the interpretation of these phenomena it would be useful to say a few words about solar activity as a whole.

## 1. Magnetic Nature of Solar Activity

There are now many arguments for believing that solar activity is of magnetic origin (see, for example, Alfvén, 1950). This is implied by the magnetic nature of active phenomena such as sunspots, active regions, flares, coronal structure etc. In general we can imagine the following sequence of phenomena as giving rise to solar activity. The magnetic field is generated by some kind of magnetohydrodynamic dynamo in the convective zone, then it is transported to the solar atmosphere. Here the magnetic field is partly dissipated, supplying one part of the energy needed for solar activity, and partly penetrates the sun's atmosphere changing the equilibrium energy transport and in this way supplying another part of the energy of solar activity.

The share of energy in solar activity is of course very feeble compared with the total energy output of the sun, which is about $4 \times 10^{33}$ erg/sec. Even for the most powerful active phenomenon – the solar flare – the energy output does not usually exceed $3 \times 10^{29}$ erg/sec. However the essential point is that the energy is delivered in forms that are almost absent for the quiet sun. These forms are just the high-temperature heating, plasma ejection, and particle acceleration, the mechanisms that are the subject of this review.

In considering the role of the magnetic field, we can distinguish three different layers in the solar atmosphere and just below. The first layer is below the photosphere where the magnetic forces are small as compared with the kinetic pressure of the plasma and the dynamic pressure of the convective motions. This is the region where the magnetic field is created and amplified by some kind of dynamo process. Here are the main sources of magnetic field which is observed in the sun's atmosphere.

The second layer includes the solar chromosphere and low corona. This is the region where for the magnetic field of sunspots and active centers the conditions

$$H^2/8\pi \gg nkT \quad \text{and} \quad H^2/8\pi \gg \tfrac{1}{2}\varrho v^2 \tag{1}$$

(i.e., magnetic pressure greatly exceeding plasma and dynamic pressures) are fulfilled.

*Dyer (ed.), Solar Terrestrial Physics / 1970: Part I*, 119–133. *All Rights Reserved.*

This region is the only one on the sun where the overall energetics and plasma behaviour can be fully controlled by the magnetic field. Because of the inequalities (1) the magnetic field must be a force-free field or, as in the highly symmetrical cases, simply a potential field.

The third region is the solar corona, some hundreds of thousand kilometers higher, where the magnetic forces are small compared with the dynamic pressure of solar wind. Here again the magnetic field becomes a minor factor.

Later we shall proceed from the assumption that flares are magnetically driven phenomena and hence we shall only consider the second range of heights in the sun's atmosphere. But we must have in mind that the origin and the main sources of the field are located below the photosphere, in the first layer.

## 2. Two Main Classes of Flare Models

In fact, almost all the hypotheses concerning the origin of flares connect the flare phenomenon with the strong magnetic field of active regions. (Note here the latest and very comprehensive review of Sweet, 1969.) These flare models can be divided in two classes. One of them comprises models for which the magnetic field plays the passive role, something like a container or trap, with the energy being stored in non-magnetic forms. These forms might be the excess gas pressure (Carmichael, 1964; Sturrock, 1968) or the pressure of accelerated protons (Elliot, 1969) being trapped in the closed magnetic field. In both cases the rupture of the field is identified with a flare. In the model of Pneuman (1968) the energy of standing waves excited in a closed magnetic field by the flux of magnetoacoustic waves from the photosphere plays the same role.

Despite a number of successes in the interpretation of separate properties of flares, all these hypotheses suffer the same defect, namely that there is no observational evidence for the accumulation of energy in the above-mentioned forms before the solar flare. Moreover, observations almost definitely show that in the flare region there is no sizable energy store before the flare except the energy of the magnetic field.

This fact, together with other arguments in favour of magnetically driven models, gives the reason for restricting ourselves to this second class of models. Thus, we shall consider the flare as a result of release of magnetic energy of the active region above the photosphere.

As for the energy store the magnetic energy is quite sufficient to produce all the observed flare phenomena. In fact this is the main argument in favour of magnetically driven models which suppose that the energy is stored in the field itself. Of course we are interested only in the excess of magnetic energy over and above the energy of purely potential magnetic field. The direct observations by Severny (1965) of strong vertical currents above active regions show that such an excess exists and is large enough. This excess of magnetic energy over the energy of potential field is an essential point for the following discussion.

Now the question arises, what is the field configuration which produces this energy excess and how does this configuration arise? The magnetically driven models can be divided into two main groups according to the answer on this question.

In the first group it is supposed that according to inequalities (1) above the magnetic field is force-free and the excess energy is contained in one (Alfvén and Carlquist, 1967; Carlqvist, 1969) or two (Gold and Hoyle, 1960) isolated force-free twisted tubes of magnetic flux, the ends of which go into the photosphere. The origin and stability of such a configuration have not been considered in detail. There are reasons to believe, however, that this configuration is unstable from the beginning and can be stabilized only by the strong potential magnetic field of surrounding space (Syrovatsky, 1969a).

So we finally come to the question of the appearance of regions of strong currents in a plasma embedded initially in the potential magnetic field. If inequalities (1) are fulfilled in the genesis of such regions it is natural to connect the deformation of the field with changes of its sources, for instance, the motion of sunspots in the photosphere or the protruding of new magnetic fields into the atmosphere from beneath the photosphere.

In general the latter process gives rise to the appearance of current sheets. This is true if there are neutral points in the initial potential magnetic field. For the simplest geometry the appearance and behaviour of current sheets were considered in a second group of papers (Giovanelli, 1946, 1947, 1948; Sweet, 1958; Dungey, 1958; Petschek, 1964; Syrovatsky, 1966, 1969a, b) which connect the flares with a strong current near the zero (neutral) line of the magnetic field. Giovanelli was the first to consider the neutral line to be the place in the solar atmosphere where the energy of a flare is delivered.

## 3. Magnetic Cumulation near the Magnetic Neutral Line

The significance of neutral lines for our problem can be illustrated by Ohm's law for a moving plasma, which we consider for simplicity to be isotropic:

$$\mathbf{j} = \sigma\left(\mathbf{E} + \frac{1}{c}[\mathbf{v} \times \mathbf{H}]\right). \tag{2}$$

Here $\sigma$ is the plasma conductivity, **E** and **H** are the electric and magnetic fields, **j** is the current density, and **v** is the bulk velocity of a plasma. It is seen from Equation (2) that because of the high conductivity ($\sigma \to \infty$) of the plasma an electric field simply produces a drift motion of the plasma with drift velocity $v = c[E \times H]/H^2$. This is true everywhere except in the vicinity of the points where $H = 0$. At these points the electric field can not be balanced by the induction and produces a strong current in the plasma.

As we shall see later, situations like this arise when sources initially producing a neutral magnetic line change in time. Such a change generally produces a uniform electric field near the neutral line which gives rise to a drift motion of the plasma outside the neutral line and a strong electric current in its immediate vicinity.

The usual approach to the problem of plasma flow and the release of magnetic energy near a neutral line is an examination of stationary flow (Sweet, 1958; Petschek, 1964; Parker, 1963). But the existence of a stationary solution is an open question. In fact, it can be shown that for some realistic initial and boundary problems the solution is in fact nonstationary and can be considered as quasistationary only in finite space and time intervals (Syrovatsky, 1966, 1969b).

To elucidate the complex process of current sheet development and behaviour we shall consider the simplest situation when the neutral line is produced by a bipolar group with equivalent magnetic moment $m$ and superimposed uniform external magnetic field $H_0$, which is parallel to the moment $m$ (see Figure 1). The field $H_0$ may be produced by neighbouring spots or a preceding active region. Note that the field topology in this case is fully equivalent to the usually considered case of a neutral line between two bipolar spot groups.

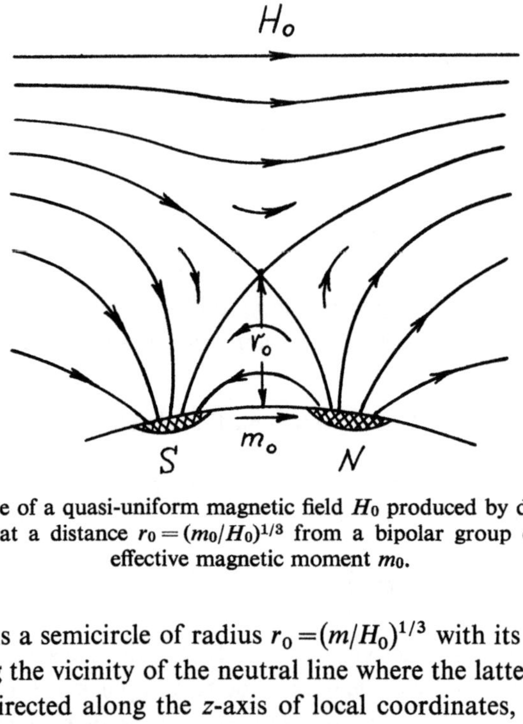


Fig. 1. In the presence of a quasi-uniform magnetic field $H_0$ produced by distant sources the zero (neutral) line appears at a distance $r_0 = (m_0/H_0)^{1/3}$ from a bipolar group of sunspots having an effective magnetic moment $m_0$.

The neutral line is a semicircle of radius $r_0 = (m/H_0)^{1/3}$ with its ends on the photosphere. Considering the vicinity of the neutral line where the latter can be considered as a straight one directed along the $z$-axis of local coordinates, we can express the electromagnetic field through the vector-potential,

$$A(x, y, t) = \tfrac{1}{2} h_0 \left(x^2 - y^2 - \delta(t)\right). \tag{3}$$

Here $h_0 = 3H_0/r_0$ and $\delta(t) = r_0^2 \Delta m(t)/3m$, where $\Delta m(t) = m(t) - m(0) \ll m(0)$ is the change of magnetic moment of the group. The expression (3) is true in the absence of plasma or in a region where

$$H^2/4\pi \gg nkT \quad \text{or} \quad r^2 \gg r_s^2 = \frac{4\pi nkT}{h_0^2}, \tag{4}$$

where $n$ is number density of the plasma and $T$ its temperature.

The change of the sources of the magnetic field produces a change of the vector-potential which propagates to the neutral line in the form of a converging cylindrical wave. This wave has a specific cumulative property: as it approaches the neutral line the amplitudes of the disturbances in velocity, field, current, and density grow without limit (Syrovatsky, 1966, 1969a, b). Because of (4) this wave propagates with nearly the local Alfvén velocity and rearranges the field in conformity with a new value of the potential (3) corresponding to the value of $\delta(t)$ (see Figure 2). It follows from

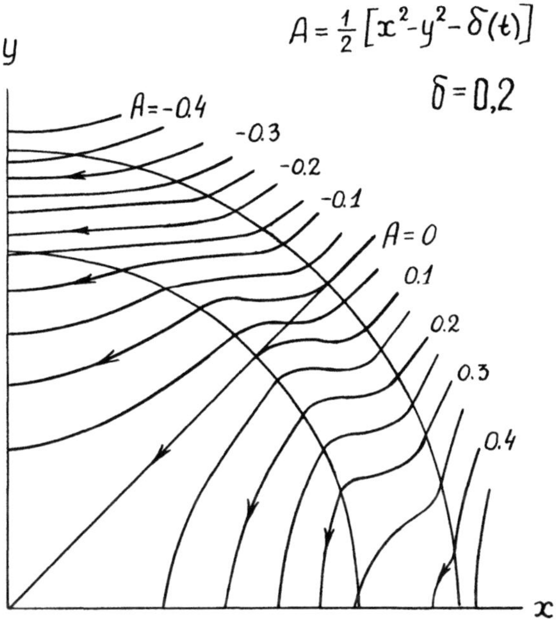


Fig. 2. A converging cylindrical wave near the neutral line which is assumed to coincide with the $z$-axis. The picture is symmetrical with respect to the $x$- and $y$-axes. The magnetic field lines and the forward and back wave fronts are shown. The potential $A$ increases by the value $-\delta/2 = -0.1$ after the passage of the wave.

Equation (3) that in the vicinity of the neutral line there is a uniform electric field along this line which equals:

$$E = -\frac{h_0}{2c}\frac{\partial \delta}{\partial t} = -\frac{h_0}{2c}\dot{\delta}. \tag{5}$$

This is the field which produces a drift motion of plasma outside the neutral line and a strong electric current near this line.

In a time of the order of

$$t_A = \frac{\sqrt{4\pi n M}}{h_0} \ln \frac{r_0}{r_s}, \tag{6}$$

the disturbance reaches the neutral line where the next stage of the process, the development of a current sheet, begins.

So for the solar atmosphere the first stage of the process lasts a few minutes (see Equation (6)) and is characterized by the beginning of magnetic field changes and plasma motions.

## 4. Current Sheet Development; Soft Phase of the Flare

Let us consider the second stage of the current sheet development. It has been recognized for many years (see Sweet, 1958; Dungey, 1958) that changes in the magnetic field outside a neutral line give rise to the formation of a current sheet on this line. The rigorous solution of the problem was given by Chapman and Kendall (1963) for an incompressible perfectly conducting fluid and by Imshennik and Syrovatsky (1967) for a compressible one. For the latter case the flow is such that the current density **j**, plasma density $n$ and specific current density $j/n$ go to infinity in a finite time. The solution is homologous and does not allow obtaining the general solution of the boundary problem. Nevertheless these homologous solutions show rigorously the property of a neutral line to develop a current sheet.

To solve the general boundary problem for a given change in the field outside the vicinity of the neutral line we must numerically integrate the full non-linear system of magnetohydrodynamic equations. An example of the results of this numerical integration was given by Syrovatsky (1969b) (see Figure 3). One of the properties of

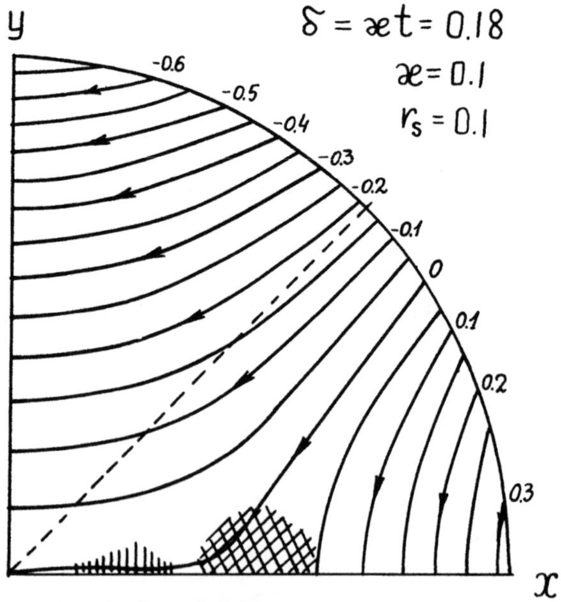


Fig. 3. Example of numerical computations of magnetohydrodynamic flow. The values of the potential $A$ at a given magnetic line of force are shown near the circumference. The regions of compression and of dilatation of plasma are shown by double and single hatching respectively.

the solution which can also be inferred from the solution of linearized equations is the width of the current sheet which is of the order of the diameter of the nonlinear region of cumulation, and is approximately given by the expression

$$x_c \cong \sqrt{\delta}\,. \tag{7}$$

In a region of this radius near the neutral line the magnetic field is nearly uniform, with the strength

$$H \cong h_0 \sqrt{\delta}\,, \tag{8}$$

and its sign changes abruptly inside the current sheet. Hence the surplus of magnetic energy over the energy of the initial potential field equals

$$\Delta W_\delta = W_\delta(t) - W_\delta(0) \approx W_\delta(0) = \frac{\pi}{16}\, h_0^2 \delta^2 r_0\,, \tag{9}$$

which is the initial potential field energy in the non-linear region of radius $\sqrt{\delta}$.

Equation (9) answers the first of the two main questions of flare theory: where and in what form does the flare energy accumulate?

Now let us consider the processes inside the current sheet. If the plasma input velocity is small compared to Alfvén velocity the current sheet can be considered quasistationary in the sense that the momentum equation must be satisfied, that is:

$$\frac{H^2}{8\pi} \cong \frac{h_0^2 \delta}{8\pi} = n_c(kT_e + kT_i) \simeq n_c kT_e\,, \tag{10}$$

where $T_e$ and $T_i$ are the electron and ion temperatures. Here we have taken $T_e \gg T_i$ because the electron energy in the current sheet acquired from the electric field is much greater than the proton energy. Denoting the sheet thickness by $y_c$ we have for the current density and the current velocity the expressions

$$j_c = \frac{c}{4\pi} \frac{h_0 \sqrt{\delta}}{y_c} \quad \text{and} \quad v_c = \frac{c}{4\pi n_c e} \frac{h_0 \sqrt{\delta}}{y_c}\,. \tag{11}$$

The sheet thickness $y_c$ is determined either by the particle gyroradius or by the finite conductivity $\sigma$ of a plasma. In the first case

$$y_c \approx \frac{c p_i}{eH} \approx \frac{c \sqrt{2MkT_i}}{e h_0 \sqrt{\delta}} \tag{12}$$

where $p_i$, $M$ and $T_i$ are the momentum, mass, and temperature of ions. But it can be easily seen in this case that the current velocity (11) will be greater than the iono-acoustic velocity

$$v_c \approx \sqrt{\frac{2kT_e}{M} \frac{T_e}{T_i}} > \sqrt{\frac{2kT_e}{M}}\,. \tag{13}$$

This is just the condition of excitement of iono-acoustic turbulence which reduces the conductivity to some finite value $\sigma$, even if it was infinite initially.

Thus the development of a current sheet leads automatically to plasma turbulence and correspondingly to a fast rate of magnetic field dissipation. The role of turbulent resistivity in Petschek's model of plasma flow near the magnetic neutral line was elucidated by Friedman and Hamberger (1969). They have shown by numerical examples that the rate of dissipation by turbulent resistivity is fast enough to produce the observed energy output in flares.

We must stress here that this result is quite general and does not depend on the value of $\sigma$. Namely, the sheet thickness if determined by finite conductivity is given by

$$y_c = \frac{c^2}{2\pi\sigma} \frac{\sqrt{\delta}}{\dot{\delta}}, \tag{14}$$

where we have used the expression $v \approx \dot{x}_c = \dot{\delta}/2\sqrt{\delta}$ for the plasma input velocity into the current sheet (see Syrovatsky, 1969a). From Equations (5), (11), and (14) we see, that Ohm's law $j = \sigma E$ is valid and that power delivered in a sheet of length $\pi r_0$ is given by

$$P = \pi r_0 y_c \sqrt{\delta} \frac{j_c^2}{\sigma} = \pi r_0 \frac{h_0^2 \delta \dot{\delta}}{8\pi}. \tag{15}$$

We see that Equation (15) does not contain the conductivity. This means simply that the sheet thickness automatically takes the value which gives the needed rate of magnetic energy dissipation. The latter is determined, as we see from Equation (15), by the value of $\delta\dot{\delta}$ and eventually by the rate of change of magnetic source configuration.

Depending on the value of $\delta\dot{\delta}$ the dissipation may vary from slow heating of some regions of the chromosphere to very powerful flares.

Hence Equation (15) at least partly answers the second main question of flare theory, which concerns the mechanism of fast release of energy in the flares. We see that development of plasma turbulence may serve as one such mechanism, giving rise to the effective resistivity of the current sheet and turbulent heating of the plasma in the sheet. We believe, however, that this answer is incomplete, indeed, the development of plasma turbulence in the sheet can be the cause of heating of plasma to high temperature and of the corresponding soft X-ray emission, of non-thermal radio noise generation and also of acceleration of some ions to high energy because of a run-away process or some stochastical mechanism. Besides, thermal conductivity along the magnetic lines of force can cause the heating of adjacent cooler regions giving rise to line emission, primarily that of the hydrogen lines. All these phenomena are really observed in flares. Moreover, for small flares these phenomena comprise almost the whole complex of the observed phenomena.

But if we consider large flares, then the existence of impulsive 'hard' phase of a flare when a large amount of energy is released in high-energy X-rays, UV, and corpuscular emission remains unaccounted for. Before going on to consider the possible mechanisms of that hard phase let us consider in more detail the property of plasma flow near the magnetic neutral line.

## 5. Distribution and Motion of Plasma near the Current Sheet

It is a very complicated problem to calculate the density and velocity distribution of plasma near the neutral line. We can obtain some insight by using the solutions of linearized equations and examples of numerical calculation of the rigorous problem. The picture is very nonuniform. Near the current sheet there exist steep gradients of the magnetic field, the plasma density, and the velocity, and here are intermittent regions of condensation and rarefaction of the plasma. Nevertheless we can distinguish some characteristic feature of the process which can be compared with observations of flares.

First of all we should like to draw attention to the very close contact of magnetic fields of opposite directions which is characteristic for the current sheet. In the case of a rising bipolar group of spots at the place of the neutral line there appears a horizontal current sheet (see Figures 1 and 4). The disk observations must show in this case the neutral line of the longitudinal magnetic field and a very close contact of oppositely directed transverse magnetic fields. It is just this property of the magnetic field in flares which seems to be typical as was stressed repeatedly by Severny (this volume, p. 38).

Furthermore, the theory predicts a surprising property of plasma flow near the zero line, namely, that there appear two filaments of dense plasma on the borders of a current sheet. These filaments are parallel to the neutral line and are located on both sides of it (see Figure 4). The filaments move away with a velocity $v \approx \dot{\delta}/2\sqrt{\delta}$ as time goes on. Their appearance is due to the squeezing of plasma out of the region $r < x_c$ near the neutral line. Simultaneously the average plasma density decreases in that region.

The appearance of two expanding filaments beside the neutral line of the magnetic field is a typical property of large chromospheric flares, as was pointed out particularly by Švestka (1962). It is interesting to note that a similar behaviour occurs for loop prominence systems, near the top of which appear two parallel rows of condensations, the plasma from which flows down into the chromosphere (Bruzek, 1964). In this respect, from the point of view of plasma dynamics, flares and loop prominence systems can be considered phenomena of the same type.

The theory of plasma motion near the neutral line of a magnetic field predicts such a two-ribbon structure in a natural way. From the known distance between ribbons and the velocity of their expansion the important parameters of the process $\delta$ and $\dot{\delta}$ may be determined. In addition we know the magnetic field gradient $h_0$, the height of flare $r_0$ and the mean plasma density $n_c$ in the current sheet, then many important parameters of flare can be estimated. These estimates for some representative conditions are given in Table I (Syrovatsky, 1969a).

As was pointed out above, the plasma density distribution in the region of cumu-

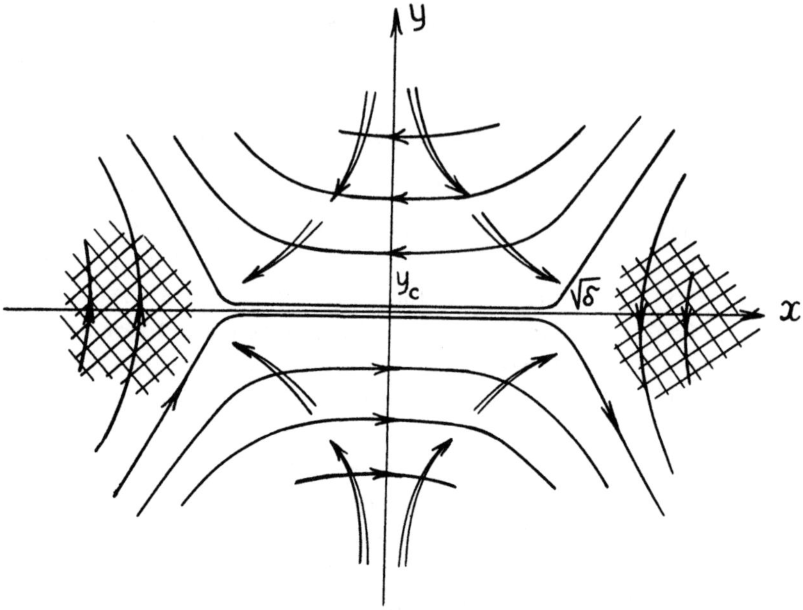


Fig. 4. The current sheet developing at the place of the initial neutral line. The sheet thickness and width are $y_c$ and $\sqrt{\delta}$ respectively. Double hatching denotes the high density plasma. Arrows show the plasma flow.

TABLE I

Parameters of flare model

| | | | | | | |
|---|---|---|---|---|---|---|
| $r_0$, cm | $10^9$ | $10^9$ | $3 \times 10^9$ | $3 \times 10^9$ | $10^{10}$ | $10^{10}$ |
| $h_0$, gauss/cm | $10^{-6}$ | $3 \times 10^{-6}$ | $3 \times 10^{-7}$ | $10^{-6}$ | $10^{-8}$ | $3 \times 10^{-8}$ |
| $H_0$, gauss | $3 \times 10^2$ | $10^3$ | $3 \times 10^2$ | $10^3$ | 30 | $10^2$ |
| $m_0$, gauss·cm$^3$ | $3 \times 10^{29}$ | $10^{30}$ | $10^{31}$ | $3 \times 10^{31}$ | $3 \times 10^{31}$ | $10^{32}$ |
| $\sqrt{\delta}$, cm | $3 \times 10^8$ | $3 \times 10^8$ | $10^9$ | $10^9$ | $3 \times 10^9$ | $3 \times 10^9$ |
| $\delta$, cm$^2$/sec | $10^{15}$ | $10^{15}$ | $3 \times 10^{15}$ | $3 \times 10^{15}$ | $10^{16}$ | $10^{16}$ |
| $V$, cm/sec | $1.5 \times 10^6$ | $1.5 \times 10^6$ | $1.5 \times 10^6$ | $1.5 \times 10^6$ | $1.5 \times 10^6$ | $1.5 \times 10^6$ |
| $n_0$, cm$^{-3}$ | $10^{12}$ | $10^{12}$ | $10^{10}$ | $10^{10}$ | $10^8$ | $10^8$ |
| $n_c$, cm$^{-3}$ | $2 \times 10^{13}$ | $2 \times 10^{13}$ | $2 \times 10^{11}$ | $2 \times 10^{11}$ | $2 \times 10^9$ | $2 \times 10^9$ |
| $n$, cm$^{-3}$ | $10^5$ | $10^5$ | $5 \times 10^3$ | $5 \times 10^3$ | $10^2$ | $10^2$ |
| $\sigma$[a], cm$^2$/sec | $10^{12}$ | $10^{12}$ | $10^{11}$ | $10^{11}$ | $10^{10}$ | $10^{10}$ |
| $y_c$, cm | 50 | 50 | $5 \times 10^2$ | $5 \times 10^2$ | $5 \times 10^3$ | $5 \times 10^3$ |
| $v_c$, cm/sec | $2 \times 10^6$ | $5 \times 10^6$ | $2 \times 10^7$ | $5 \times 10^7$ | $2 \times 10^7$ | $5 \times 10^7$ |
| $T_e$, K | $1.5 \times 10^6$ | $1.5 \times 10^7$ | $1.5 \times 10^8$ | $1.5 \times 10^9$ | $1.5 \times 10^8$ | $1.5 \times 10^9$ |
| $\Delta W_\delta$, erg | $2 \times 10^{30}$ | $2 \times 10^{31}$ | $6 \times 10^{31}$ | $6 \times 10^{32}$ | $2 \times 10^{31}$ | $2 \times 10^{32}$ |
| $P$, erg/sec | $10^{28}$ | $10^{29}$ | $10^{29}$ | $10^{30}$ | $10^{28}$ | $10^{29}$ |
| $t = m/\dot{m}$, sec | $2 \times 10^2$ | $3 \times 10^2$ | $10^3$ | $10^3$ | $3 \times 10^3$ | $3 \times 10^3$ |
| $V_c$[b], cm$^3$ | $10^{20}$ | $10^{20}$ | $10^{22}$ | $10^{22}$ | $10^{24}$ | $10^{24}$ |
| $N_c = n_c V_c$ | $2 \times 10^{33}$ | $2 \times 10^{33}$ | $2 \times 10^{33}$ | $2 \times 10^{33}$ | $2 \times 10^{33}$ | $2 \times 10^{33}$ |
| $n_c^2 V_c$, cm$^{-3}$ | $4 \times 10^{46}$ | $4 \times 10^{46}$ | $4 \times 10^{44}$ | $4 \times 10^{44}$ | $4 \times 10^{42}$ | $4 \times 10^{42}$ |

[a] In the estimates Buneman's value of the conductivity was used: $\sigma = \frac{1}{2}(M/m)^{1/3}\,\omega_{0c} \approx 3 \times 10^5 \sqrt{n_c}$.

[b] $V_c$ is the volume of the current sheet.

lation $r < x_c$ is highly non-uniform. Here only very crude and not very reliable estimates can be made. On the average for the current sheet the plasma density increases to approximately

$$n_c \approx n_0 \ln\left(\sqrt{\delta}/y_c\right) \tag{16}$$

(here $n_0$ is the initial uniform plasma number density). At some points in the sheet, however, it can attain a very large value, for instance at $r=0$ (Sweet, 1969). Outside the current sheet but adjacent to it, the plasma density may have a very low value because of the squeezing of plasma to the sheet borders; thus a thin sheet of plasma carrying a large current is surrounded by a rarefied plasma. As we shall see later, the instability of such a sheet with respect to tearing may be the cause of the hard phase of the flare.

Having in mind the observed high-velocity plasma ejection during flares, we need to consider the plasma velocity distribution near the current sheet. A numerical calculation shows that the plasma velocity near the current sheet can reach a value of the order of the Alfvén velocity which exceeds the thermal velocity of the plasma by a large amount. The plasma flows with a velocity of this order along the current sheet and is thrown out into the dense filaments at the border of the sheet (the filaments themselves move with a much smaller velocity). This plasma flow may be the cause of the spectroscopically observed high velocities in flares. Especially large velocities of plasma ejections can be achieved when the sheet ruptures, giving a 'slingshot' or 'melon seed' effect. In this case plasma is ejected along the sheet and then along the magnetic field lines emerging from the sheet borders. In this model these impulsive high-velocity ejections of plasma can be considered the beginning of the 'hard' phase.

## 6. Current Sheet Rupture and Hard Phase of a Flare

We have considered the 'soft' phase of a flare where there appear current sheet turbulence, high-temperature heating of electrons, electromagnetic plasma noise, and some number of run-away particles, mainly protons and heavier nuclei.

The impulsive 'hard' phase, when hard electromagnetic radiation and high energy particles appear, is of great interest. We believe that this phase corresponds to the transition from the quasi-stationary picture considered above to a highly non-stationary one due to the rupture of some part of the current sheet and interruption of the local current.

For a thin current sheet surrounded by a rarefied plasma the current break-off can occur because the current velocity reaches its limiting value $v_c \to c$ in that part of the sheet where the plasma density is much lower than its mean value (16) (Syrovatsky, 1966) or because a current sheet instability, for instance, of the tearing-mode type occurs (Sturrock, 1968). The break in the sheet corresponds to the local current interruption. As the result of this interruption there appears a strong electric field of electromagnetic (Syrovatsky, 1967) or of electrostatic (Alfvén and Carlquist, 1967) origin. In our example of current sheet break-off, the field which appears has an electromagnetic origin and can greatly exceed the quasistationary field $E$ (see Equation (5)).

The origin of this electric field can be illustrated by a simple model. Let us consider a uniform magnetic field $H_c$ directed along $x$-axis which changes sign in the plane $y=0$. We consider this plane to be the current sheet which disappears at the moment $t=0$. Then the simple solution of electrodynamic equations shows that at the moment $t=0$ an electromagnetic wave appears which transforms the magnetic field $H_c$ into the electric field

$$E = H_c\{\theta(y + ct) - \theta(y - ct)\}, \tag{17}$$

where $\theta(\xi)$ is a step function (see Figure 5).

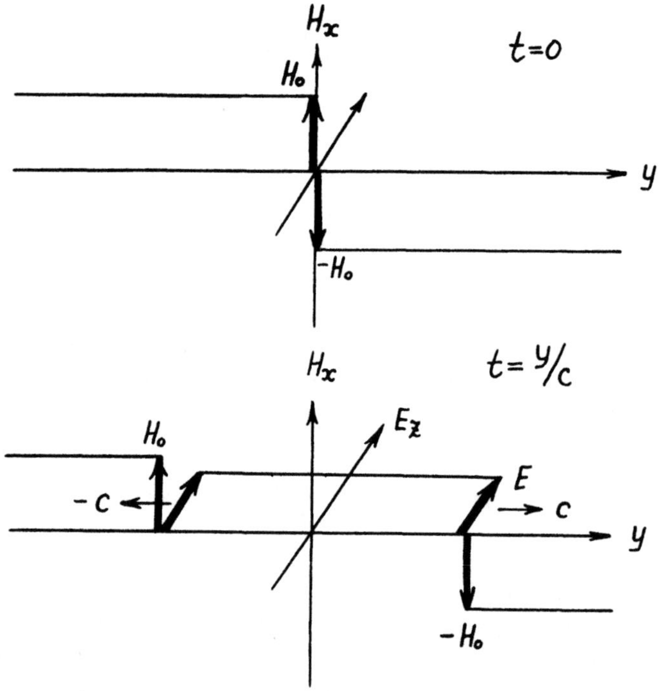


Fig. 5. Instantaneous disappearance of the current sheet in the plane $y=0$ in vacuum produces an electromagnetic wave which converts the uniform magnetic field $H_0$ into a uniform electric field $E=H_0$.

Of course this simple solution is true only in a vacuum and, what is still more important, in the case when the sheet disappearance is instantaneous. Because the sheet break-off can proceed only with a finite velocity $V$, the electric field that appears must be weaker than the field (17) and approximately equal to

$$E \approx (V/c)\, H_c. \tag{18}$$

Nevertheless, if the condition $V \gg \delta/2\sqrt{\delta}$ is satisfied, the field (18) will be much greater than the initial field (5).

For quantitative estimates concerning this impulsive hard phase, we need to specify in detail a very complicated problem in plasma kinetics. This problem is not solved yet; in fact, it is not even strictly formulated. Nevertheless the appearance of an impulsive electric field resulting from a rupture of a current sheet is a very probable hypothesis which helps one to understand the impulsive acceleration of a great number of charged particles to high energy during the short, hard phase of the flare. The observed charge composition of solar cosmic rays and the acceleration of a great number of relativistic electrons give evidence in favour of acceleration by just such an impulsive electric field, which accelerates all particles to the same rigidity irrespective of their charge and mass.

Now let us consider briefly some consequences of the flare model considered.

First of all some words about the origin of the observed flare emission in hydrogen lines. For the model under consideration, it is natural to connect the line emission with the flow along the magnetic lines of the heat from the current sheet during the soft phase and of the accelerated particles during the hard phase.

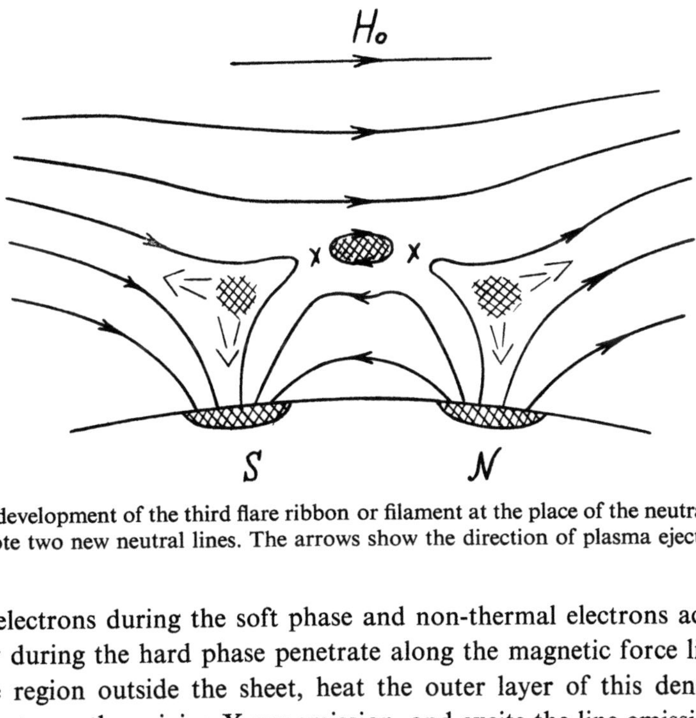


Fig. 6. The development of the third flare ribbon or filament at the place of the neutral line. Crosses denote two new neutral lines. The arrows show the direction of plasma ejections.

The hot electrons during the soft phase and non-thermal electrons accelerated to high energy during the hard phase penetrate along the magnetic force lines into the more dense region outside the sheet, heat the outer layer of this dense region to high temperatures, thus giving X-ray emission, and excite the line emission in deeper layers.

The main part of the magnetic energy which is accumulated in the region of the flare is dissipated in the course of the hard phase when it is supplied to the accelerated particles. Hence a flare may be considered as being primarily the particle event and the optical, radio and X-ray emissions are secondary processes due to these accelerated particles.

Accordingly the observed flare ribbons are probably the dense flare filaments heated by the flux of energetic particles. Two such dense filaments are usually observed on both sides of the neutral line. Sometimes, however, between two such filaments a third filament is observed (Křivský, 1969). The appearance of the third ribbon can be interpreted as a current sheet rupture at two points symmetrically disposed with respect to the neutral line (see Figure 6). In this case the plasma between these points is compressed by concentric magnetic field lines and produces the filament along the initial magnetic neutral line.

This mechanism may also be responsible for the appearance of dark filaments which are often observed along the magnetic neutral line.

## 7. Conclusions

The consideration of known flare models shows that the more promising models are those which connect the flare with the release of magnetic energy in excess of the energy of the potential field in the active region. Among these models the model of nonstationary cumulative flow near a magnetic neutral line explains the greatest number of observed facts. This model allows one to explain the main features of the soft phase of a flare and points the way towards an interpretation of the explosive hard phase.

Of course this model greatly simplifies the real situation in solar flares and even with this simplification it presents great difficulties in a quantitative development of the theory. These difficulties force us to restrict ourselves to simple crude estimates and qualitative considerations.

Nevertheless we believe that the model describes the main physical processes in flares and that it can serve as a base for a more comprehensive and, essentially, for a more precise quantitative theory.

We do not intend to discuss here the theoretical aspects which need further development and specification. They seem to be clear from the above.

From the observational point of view the questions that need further investigation are, for instance, the following:

(1) The correlation between the energy emitted by flares and the value and rate of change of the magnetic moment of developing groups of sunspots.

(2) The structure of the magnetic field in the flare region, in particular the configuration and dimensions of the current sheet.

(3) The position of the flare ribbons, their motion and relation to the neutral line.

(4) The secondary character of the optical flare with respect to the turbulent heating and acceleration of particles.

(5) The two phases of the flare, their time history and relation to other observed phenomena.

(6) The appearance of high-velocity plasma ejections and hard radiation which indicate the start of the hard phase.

## References

Alfvén, H.: 1950, *Cosmical Electrodynamics*, Univ. of Cambridge Press, Cambridge.
Alfvén, H. and Carlquist, P.: 1967, *Solar Phys.* **1**, 220.
Bruzek, A.: 1964, *Astrophys. J.* **140**, 746.
Carlquist, P.: 1969, *Solar Phys.* **7**, 377.
Carmichael, H.: 1964, in W. H. Hess (ed.), *AAS-NASA Symp. Phys. of Solar Flares*, Washington.
Chapman, S. and Kendall, P. C.: 1963, *Proc. Roy. Soc.* **A271**, 435.
Dungey, J. W.: 1958, *Cosmic Electrodynamics*, Univ. of Cambridge Press, Cambridge.
Elliot, H.: 1969, in C. de Jager and Z. Švestka (eds.), 'Solar Flares and Space Research', *XIth Cospar Symp.*, North-Holland Publ. Co., Amsterdam, p. 356.
Friedman, M. and Hamberger, S. M.: 1969, *Solar Phys.* **8**, 104.
Giovanelli, R. G.: 1946, *Nature* **158**, 81.
Giovanelli, R. G.: 1947, *Monthly Notices Roy. Astron. Soc.* **107**, 338.
Giovanelli, R. G.: 1948, *Monthly Notices Roy. Astron. Soc.* **108**, 163.
Gold, T. and Hoyle, F.: 1960, *Monthly Notices Roy. Astron. Soc.* **120**, 89.
Imshennik, V. S. and Syrovatsky, S. I.: 1967, *Zh. Exper. Teoret. Fisiki* **52**, 990.
Křivský, L.: 1969, *Bull. Astron. Inst. Czech.* **20**, 163.
Parker, E. N.: 1963, *Astrophys. J. Suppl.* **8**, 177.
Petschek, H. E.: 1964, in W. N. Hess (ed.), *AAS-NASA Symp. Phys. of Solar Flares*, Washington, p. 425.
Pneuman, G. W.: 1968, *Solar Phys.* **2**, 462.
Severny, A. B.: 1965, *Astron. Zh.* **42**, 217.
Sturrock, P. A.: 1968, in K. O. Kiepenheuer (ed.), 'Structure and Development of Solar Active Regions', *IAU Symp.* **35**, 471.
Švestka, Z.: 1962, *Bull. Astron. Inst. Czech.* **13**, 190.
Sweet, P. A.: 1958, *Nuovo Cim. Suppl.* **8**, 188.
Sweet, P. A.: 1969, *Ann. Rev. Astron. Astrophys.* **7**, 149.
Syrovatsky, S. I.: 1966, *Astron. Zh.* **43**, 340.
Syrovatsky, S. I.: 1967, *Izvestia Akademii Nauk SSSR. ser. fis.* **31**, 1303.
Syrovatsky, S. I.: 1969a, *Trudy Mezhdunarodnogo Seminara*, Leningrad, p. 7.
Syrovatsky, S. I.: 1969b, in C. de Jager and Z. Švestka (eds.), 'Solar Flares and Space Research', *XIth Cospar Symposium*, North-Holland Publ. Co., Amsterdam, p. 346.

# PARTICLE DIFFUSION IN THE SOLAR CORONA

H. ELLIOT
*Imperial College, London, England*

**Abstract.** The configuration of magnetic fields in the corona is discussed in relation to the escape of energetic solar protons into interplanetary space. In particular an assessment is made of the relative importance of diffusion, drift motions and direct field line connections in the distribution of energetic particles in the corona. Observational evidence on particle flux distributions near the earth is related to the question of storage in the corona. The absence of marked anisotropies at late times during impulsive solar events is indicative of storage in interplanetary space rather than in the corona. It is concluded that the geometry of coronal magnetic fields is such as to permit storage of protons in the energy rage 10–100 MeV for several days and that, on occasion, the fields are of such a form as to permit rapid spreading in latitude and longitude and that this, coupled with the stochastic wandering of field lines, can readily account for the observed spread in solar longitude of particles at IAU.

## 1. Introduction

The propagation of energetic solar flare particles from active regions on the sun outwards through the solar wind is a process of great complexity and any discussion of particle motion in the corona is intimately bound up with this wider aspect. The complexity is due in part to the complicated and time-dependent structure of the interplanetary magnetic field and in part to the wide spectrum of particle energies that has to be considered. At the orbit of earth, for example, where the field strength is typically $5 \times 10^{-5}$ gauss, the Larmor radius for a 40 keV electron is about $10^7$ cm, whereas, that for a 15 GeV proton is about $10^{12}$ cm. Clearly, the propagation characteristics for these two species of particles are going to be very different. The guiding centre for a 40 keV electron cannot stray very far from the line of force on which it first found itself at the solar surface, whereas, gradient drifts and scattering in field irregularities are likely to dominate the motion of the 10 GeV proton. Thus, in such a situation, some particles propagate under conditions of strictly adiabatic motion, whilst the motion of others is more appropriately treated using the concepts of anisotropic diffusion. Field line wandering caused by photospheric motions is likely to be of major importance in distributing low energy particles generated in a localized source over a wide range of solar latitude and longitude (Jokipii and Parker, 1969). Adiabatic cooling in the expanding solar wind will be significant for particles with long transit lines, whereas, for others it will be negligible (Fisk and Axford, 1968).

Although some measurements have been made of magnetic fields in prominences (Zirin and Severny, 1961) there is at the present time no way of directly measuring the general magnetic fields in the corona. Consequently the complication of the propagation process through the interplanetary magnetic field adds obscurity to ignorance and makes for very great difficulty in deducing anything about detailed behaviour of particles close to the sun, particularly as we have in any case information about the interplanetary magnetic field only in the neighbourhood of the earth's orbit. It is unlikely that this situation will change very much in the foreseeable future, but

*Dyer (ed.), Solar Terrestrial Physics / 1970: Part I,* 134–150. 

nevertheless, we must try to understand the propagation process in sufficient detail to answer some or all of the following questions because they relate to other wider issues.

(a) Do solar particles remain trapped in the corona for appreciable times prior to being released into interplanetary space?

(b) Are they accelerated on open or closed field lines?

(c) Are particles accelerated other than at times of solar flares?

(d) Do they lose a significant fraction of their initial energy as a result of the expansion of the solar wind?

(e) If there is an appreciable loss of energy, what fraction of the total flare energy did the particles carry at the point of production?

Answers to these questions relate to the basic flare mechanism, to the nature of the acceleration process and would, at the same time, allow a better estimate to be made of the stellar contribution to the low energy end of the cosmic ray spectrum – an important quantity in the general astrophysical context. In preparing this short review it seemed better to try to assess the extent to which answers to these questions are available rather than attempt to summarize the voluminous literature on solar fields and energetic particles.

## 2. Magnetic Fields in the Corona

Photospheric magnetic fields can now be measured on a routine basis by making use of the Zeeman splitting of absorption lines and it appears that there are virtually no parts of the solar surface that are entirely field-free. The field strength in the photosphere seems to be generally in the order of 1 to 10 gauss increasing to several thousand gauss in the umbrae of sunspots. The interplanetary magnetic field strength measured at or near the orbit of the earth is consistent with root fields of a few gauss at the surface of origin of the solar wind corresponding to the generally observed values in the photosphere. The much higher sunspot fields apparently do not contribute directly to the interplanetary field (Wilcox *et al.*, 1969). This is to be expected, since the field strength in the spots is sufficiently high to inhibit the coronal expansion into interplanetary space.

Whilst there is no means of directly measuring the magnetic field in the inner corona substantial progress has been made recently in computing the field configuration (Newkirk, 1967; Altschuler and Newkirk, 1969; Schatten *et al.*, 1969). Broadly speaking, these computations are based on the measured line of sight component of the photospheric field and the assumption that $\nabla \times B = 0$ between the photosphere and a concentric spherical surface of radius $R_W$. Beyond this surface the solar wind energy density dominates the situation and the field is radial. Figure 1 shows the representation of the field and plasma energy densities as a function of height above the photosphere used by Schatten *et al.* Figure 2a shows the field configuration computed for the epoch of the November 12th 1966 eclipse, together with a photograph of the corona taken during that eclipse (Schatten, 1968). Figure 2b shows a similar comparison by Altschuler and Newkirk (1969). The similarity between the geometry of the field and the coronal shapes is impressive. Similar good agreement was obtained for the September 1968 eclipse (Schatten, 1969).

## 3. Storage of Energetic Particles in the Corona

The diffusion of energetic solar flare particles in a random field in the corona has been discussed by Reid (Reid, 1964) and by Axford (Axford, 1965) using a model shown schematically in Figure 3. Diffusion in the thin shell concentric with the sun is

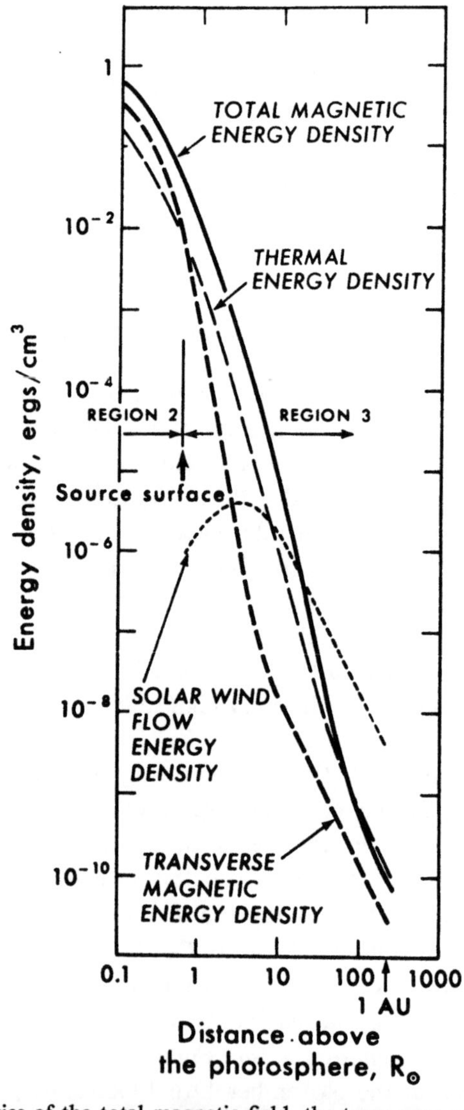


Fig. 1. The energy densities of the total magnetic field, the transverse magnetic field, the thermal motions and the solar wind flow as a function of height above the photosphere (after Schatten *et al.*, 1969).

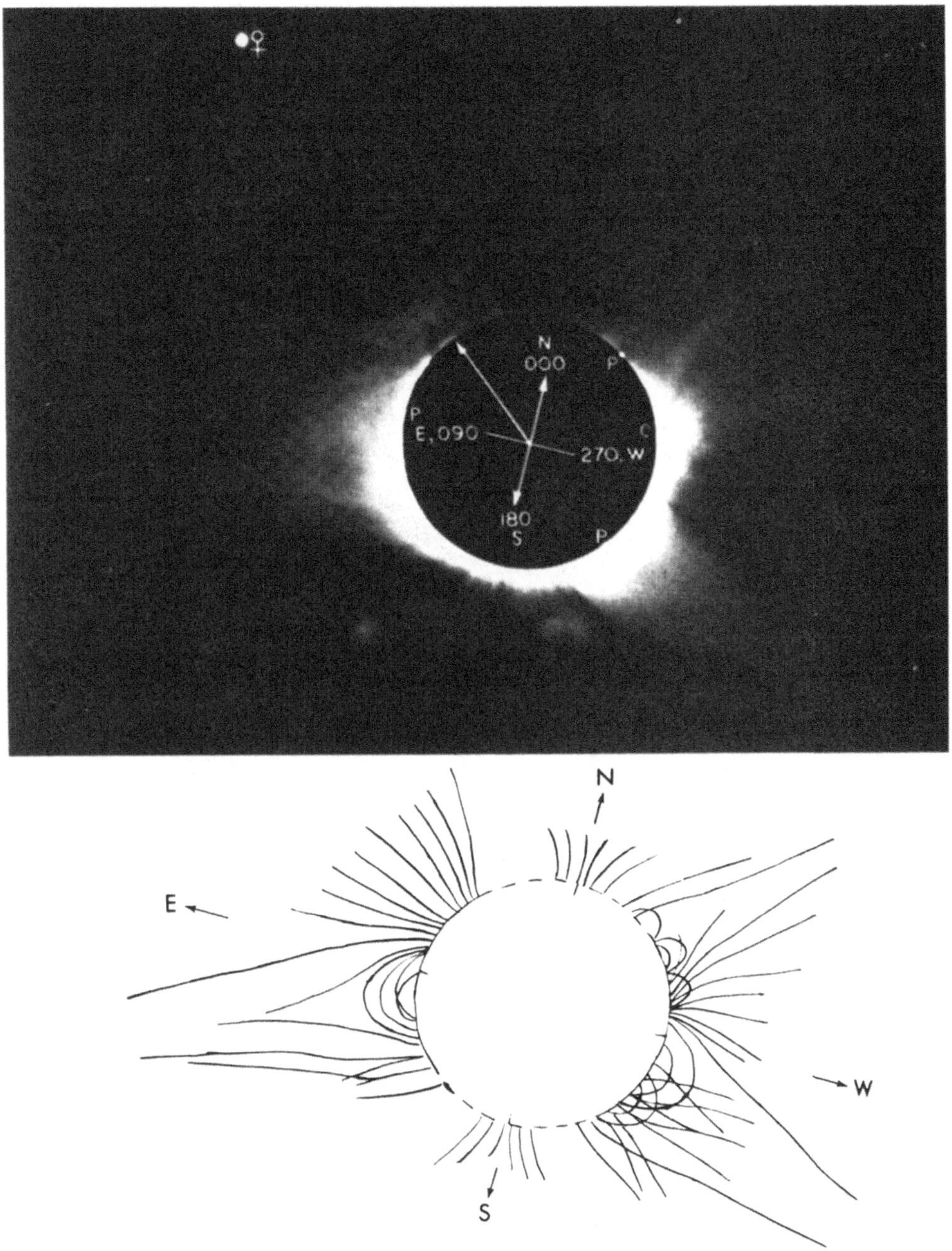


Fig. 2a. Comparison between the optical corona and computed coronal fields for the November 12th 1966 eclipse (after Schatten 1968).

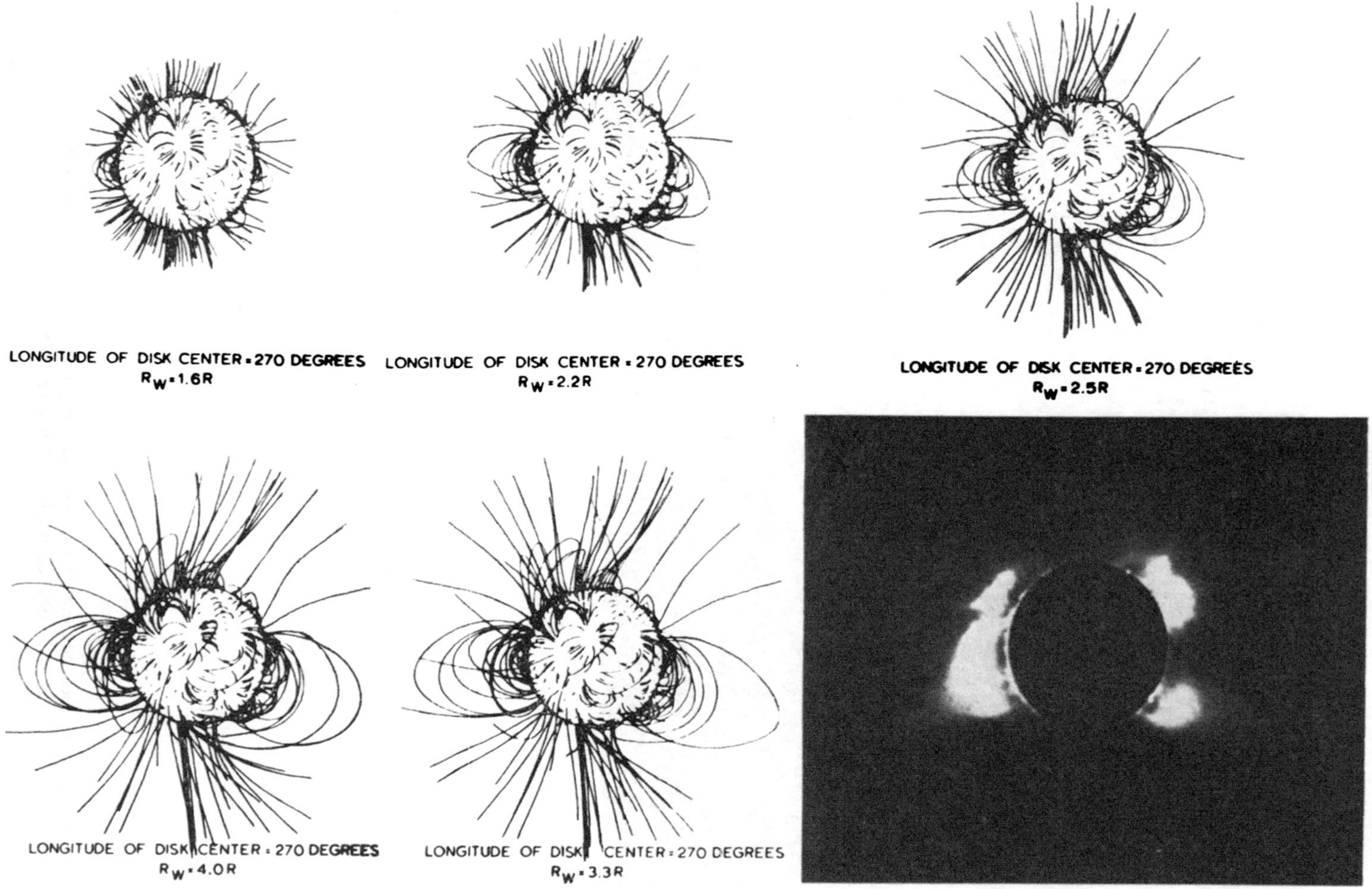


Fig. 2b. Comparison between the optical corona and computed coronal fields for the November 12th 1966 eclipse (after Altschuler and Newkirk 1969).

taken to be isotropic, whereas, that along the spiral field of the solar wind is assumed to be highly anisotropic. The thin layer of highly disordered magnetic field used by Reid and Axford in their model is clearly a poor representation of the field patterns shown in Figure 2 and any agreement between the observed solar proton behaviour

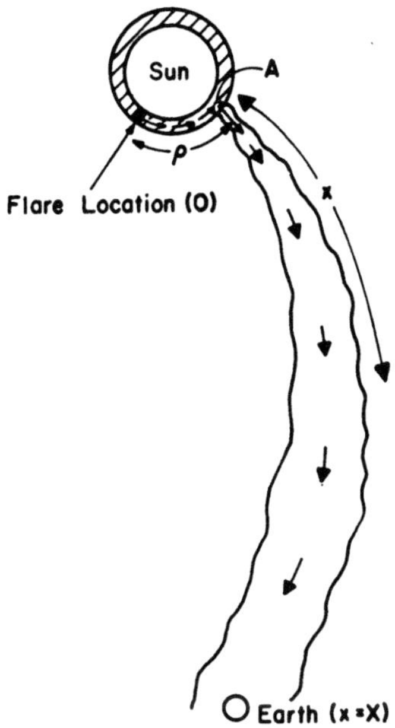


Fig. 3. Configuration of coronal and interplanetary magnetic fields used for diffusion calculation by Reid (1964) and Axford (1965). The arrows indicate the path of protons which eventually reach the earth.

and the predictions of the model serves only to show that the morphology of solar proton events measured far from the sun is decidedly insensitive to details of the coronal field configuration.

The existence of closed loops in the calculated field patterns clearly provide the kind of geometry most favourable to particle storage. The field strengths in the coronal loops are likely to be typically a few tens of gauss over a large bi-polar spot group increasing, of course, with decreasing height and quite consistent with the values of a hundred or so gauss deduced by Holt and Ramaty (Holt and Ramaty, 1969) for their model for the production of microwave and X-ray bursts from solar flares.

In such a field the Larmor radius for the highest energy solar particles is a few kilometres only. The escape of such particles could conceivably be determined either by scattering in field irregularities or by drift motion due to field gradients and curvature of field lines. If we take $L$ as the scale size of the field and $R_L$ as the Larmor radius for a particle with velocity $v$ then we may make very rough estimates of the particle transport times appropriate to these three processes as follows:

1. *Scattering*

If we assume that the centre of gyrations random walks with a step length $R_L$ each gyro period then the time required for the guiding centre to move a distance $L$ perpendicular to the field is given by

$$\tau_s \approx \frac{18\pi}{v_\perp} \frac{L^2}{R_L}.$$

2. *Curvature Drift*

Take the radius curvature $\approx$ the scale size $L$, then

$$\tau_c \approx \frac{v_\perp}{v_\parallel^2} \frac{L^2}{R_L}.$$

3. *Gradient Drift*

Take the field gradient to be $\approx B/L$. Then

$$\tau_G \approx \frac{2}{v_\perp} \frac{L^2}{R_L}.$$

Taking $2 \times 10^{10}$ cm for the scale size $L$ and 10 gauss for the field strength we can estimate $\tau_c$, $\tau_G$ and $\tau_s$. Order of magnitude values for these times are shown in Table I.

The corresponding transfer time $\tau_p$, due to the convective motions in the photosphere carrying the feet of the field lines around is $7 \times 10^6$ sec, whilst the time for the protons to travel a distance $L$ *along* the field, $\tau_\parallel$, is in the range 1 to 10 sec. The transit times $\tau_T$ from sun to earth through the spiral interplanetary field are also shown in the table. In estimating the sun-earth travel time the path length has been taken to be $2 \times 10^{13}$ cm. An upper limit to the storage time for protons in the corona is set by energy loss due to collisions with the coronal hydrogen. This can be quite short at low energies. If, for example, we take the density of protons and electrons in the coronal region, in which trapping takes place, to be $10^8$ cm$^{-3}$ the lifetime of a 1 MeV proton is in the order of $10^3$ sec.

Figure 4 summarizes these order of magnitude estimates of characteristic times and we see that if the assumptions are correct we would expect coronal storage times to be greatest for protons in the energy range 10 to 100 MeV. In this connection it is of interest to note the suggestion (Elliot 1964, 1969) that the flare energy may be stored in the coronal magnetic fields in the form of protons of this energy which have acquired their energy in a relatively slow acceleration process culminating in the catastrophic release of most of the particles into the chromosphere and photosphere at the time of the flare. According to this picture, the flare is the *result* of proton acceleration rather than vice-versa. The storage times given here would not be incompatible with such a model.

TABLE I

Characteristic transport times for solar protons

| Proton energy (eV) | $R_L$ (cm) | $v$ (cm sec$^{-1}$) | $\tau_{\parallel}$ (sec) | $\tau_c$ (sec) | $\tau_G$ (sec) | $\tau_s$ (sec) | $\tau_p$ (sec) | $\tau_T$ (sec) |
|---|---|---|---|---|---|---|---|---|
| $3\times10^9$ | $1\times10^6$ | $3\times10^{10}$ | 1 | $1\times10^4$ | $3\times10^4$ | $8\times10^5$ | $7\times10^6$ | $7\times10^2$ |
| $3\times10^7$ | $8\times10^4$ | $8.7\times10^9$ | 3 | $6\times10^5$ | $1\times10^6$ | $4\times10^7$ | $7\times10^6$ | $3\times10^3$ |
| $10^6$ | $1.4\times10^4$ | $1.4\times10^9$ | 10 | $10^7$ | $10^7$ | $10^7$ | $7\times10^6$ | $1.4\times10^4$ |

Energetic particles travelling along the field lines of magnetic arches in the corona can clearly travel distances corresponding to a fraction of a solar radius in a matter of seconds. Curvature and gradient drifts permit the dispersion of such particles over similar distances but with the longer time scales indicated in Figure 4. An indication

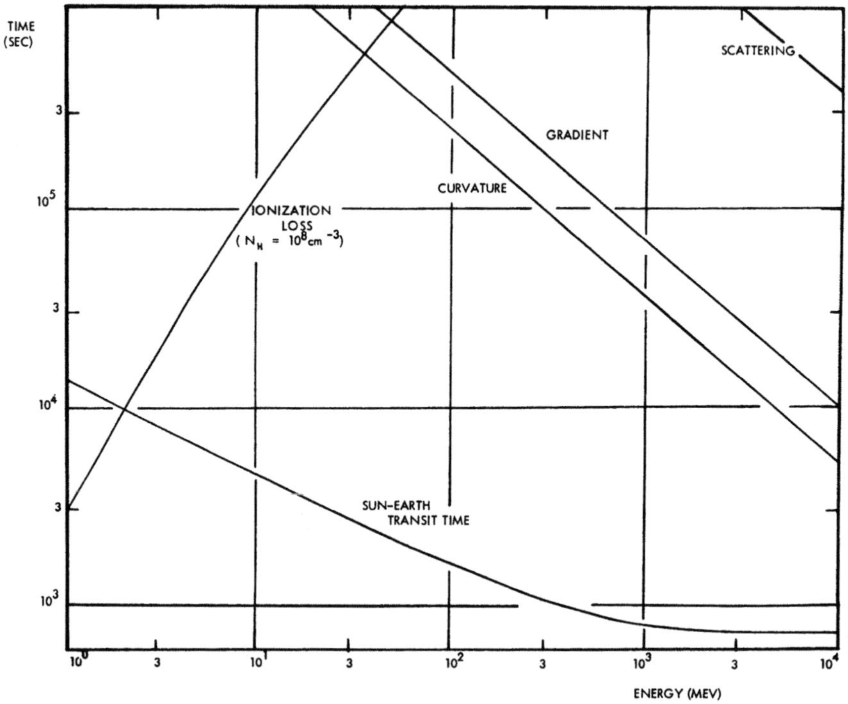


Fig. 4. Estimated characteristic transport times for solar protons undergoing scattering and curvature and gradient drifts. Also shown are the sun-earth transit time for a spiral field line and the lifetime in the corona against collision loss.

of what could be a rapid transfer of electrons in the corona has recently been reported by Tanaka and Énomé (1970). Figure 5, taken from their paper, shows a scan of the sun at a wavelength of 8 cm on January 18, 1969. The solid line shows the emission prior to the flare associated burst of radiation which is shown at different stages of development by the dotted lines. There were two distinct sources of radiation in the burst designated $R_1$ and $R_2$. The position of $R_1$ corresponded to the position of the flare over the sunspot $S_1$. $R_2$ was located at or near sunspot $S_2$ and there is no record of a flare in that position. The distance between $S_1$ and $S_2$ was $\approx 1.6 \times 10^{10}$ cm and the delay in the development of $R_2$ relative to $R_1$ was at the most a few seconds corresponding to a velocity of propagation of $\approx 1.5 \times 10^9$ cm sec$^{-1}$. This propagation velocity could conceivably be ascribed to a hydromagnetic wave but is more plausibly explained in terms of electron transit time from $S_1$ to $S_2$.

## 4. Energetic Solar Particle Events

There is now a vast accumulation of observational data on energetic solar particle events dating from the early work with ionization chambers (Berry and Hess, 1942), (Lange and Forbush, 1942; Ehmert, 1948) and with neutron monitors (Adams and Braddick, 1950). To the ground level data there has now been added a wealth of detail on low energy events derived from spacecraft. (See, for example, Fan *et al.*, 1968; Lanzerotti, 1969; Anderson, 1969; Vernov *et al.*, 1969; Cline and McDonald,

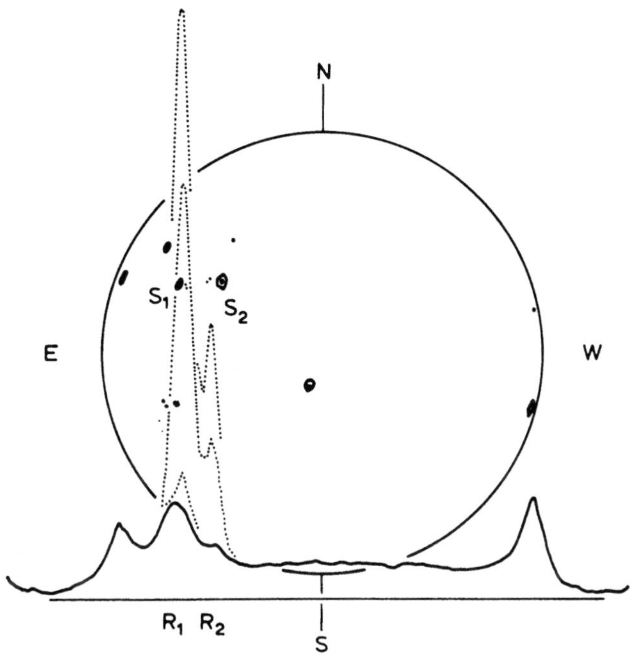


Fig. 5. One-dimensional brightness distribution at 3.75 GHz superimposed on a sunspot sketch for January 18, 1969. $R_1$ and $R_2$ represent the noise sources and $S_1$ and $S_2$ the corresponding sunspots (after Tanaka and Énomé, 1970).

1968.) It is not the intention here to attempt to review this extensive collection of data but rather to select examples that appear to be relevant to the questions posed in Section 1.

Figure 6 shows the first two flare events to be recorded (Lange and Forbush, 1942). The absence of any appreciable increase in particle flux in the equatorial region indicates that very few, if any, of the solar particles had energy in excess of $10^{10}$ eV on these two occasions. At high latitudes the solar particles continue to arrive for a period of up to 12 hours – much longer than the duration of the optical flares.

Figure 7 shows the flux of solar protons with $0.6 < E < 13$ MeV registered by detectors on board Pioneer 6 (Fan *et al.*, 1968) together with neutron monitor data from Climax. In this figure we see two examples of solar proton events which in each case span a period of approximately 10 days and which consist of a quasi-steady flux of particles occupying a rather well defined sector of the interplanetary magnetic field co-rotating with the sun. Slowly varying, co-rotating fluxes of this kind were first observed by Bryant *et al.* (1962) and can be identified by the timing of their onset at

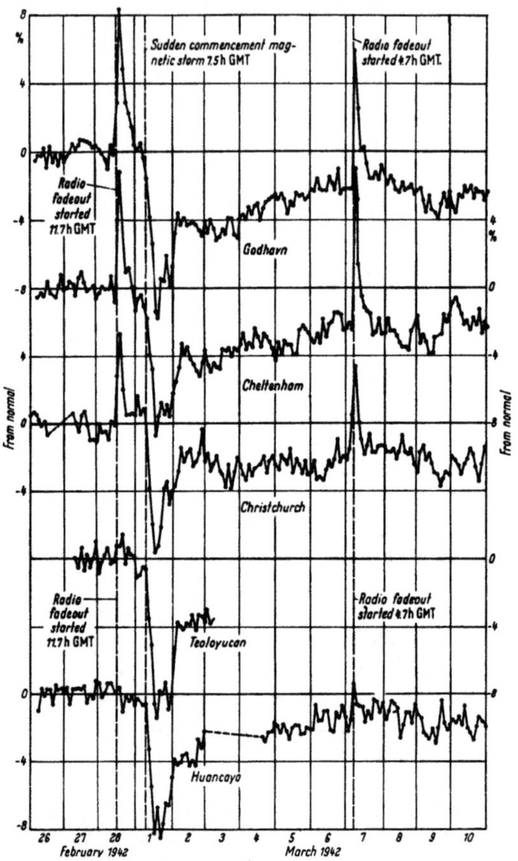


Fig. 6. Solar cosmic ray events of February 28 and March 7, 1942 registered by the Carnegie Institute ionization chambers (after Lange and Forbush, 1942).

widely separated spacecraft. Upon this slowly varying background there is superimposed a number of relatively sharply defined and discrete events, most clearly discerned at higher energies, and which can be related to specific flares. In general the duration of the impulsive events is greatest for low energy particles and decreases with increasing particle energy.

On the basis of these and other similar observations, Fan *et al.* conclude that protons with $13<E<70$ MeV generated in flares and appearing as discrete events have access to interplanetary field lines over a range of some 100° of solar longitude and that for protons with $0.6<E<13$ MeV the corresponding range is more nearly

180° roughly centred on the 100° of the higher energy discrete events. Anderson (1969) arrives at similar conclusion for solar electrons with energy 20 keV. Some of this spread in longitude presumably arises from the stochastic wandering of field lines as suggested by Jokipii and Parker (1969) but it is clear that the relative ease with which particles can move along the loops in the coronal field must play an important part in the process.

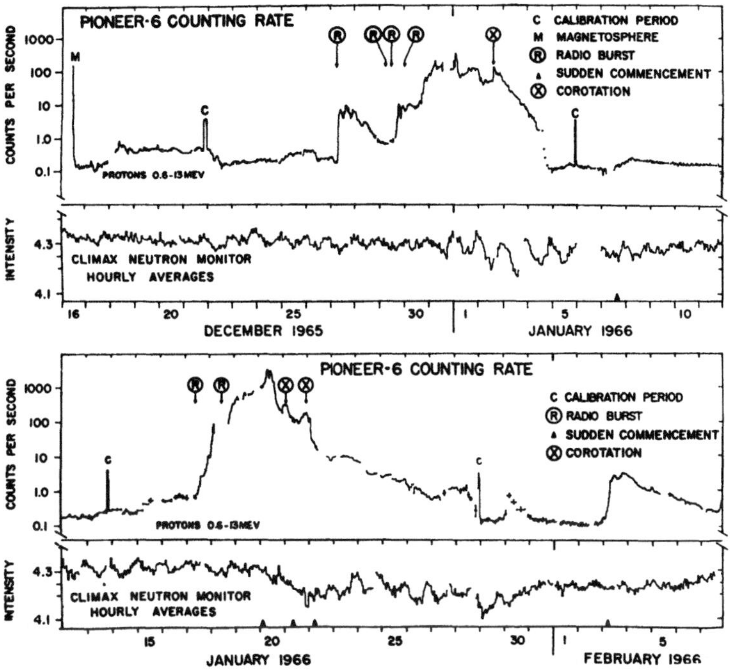


Fig. 7. Solar protons in the energy range 0.6–13 MeV. The count rates have been arranged over 30-min intervals, except for data points with error bars. Climax neutron monitor data are shown for comparison (after Fan *et al.*, 1968).

### 4.1. Delay in the onset time for impulsive events

In principle, the observed delay between the arrival of solar particles and the flare as observed at optical, radio or X-ray wavelengths should provide some information on particle storage times in the corona. Fan *et al.* (1968) observed delays of 110 min for 70 MeV protons and 200 min for 13 MeV protons emitted by a flare at 30°E solar longitude. The times for fastest transmission along the spiral field line from the western hemisphere of the sun are 25 min and 60 min respectively. This would allow storage times in the corona of up to 85 and 140 min.

Figure 8 shows some recent data from the ESRO Satellite HEOS 1 relating to the flare of February 25th, 1969 (Balogh and Hynds, 1970). This flare was situated at 38° W 12° N and according to these authors there was a delay of only $60 \pm 10$ min between flare maximum and the arrival of the earliest protons with $1 < E < 13$ MeV at the HEOS spacecraft. This corresponds to direct transmission along the interplanetary spiral field line connecting the spacecraft to the near neighbourhood of the flare and on this occasion, apparently, the maximum delay time due to storage of these particles in the corona could not have been greater than about 10 min.

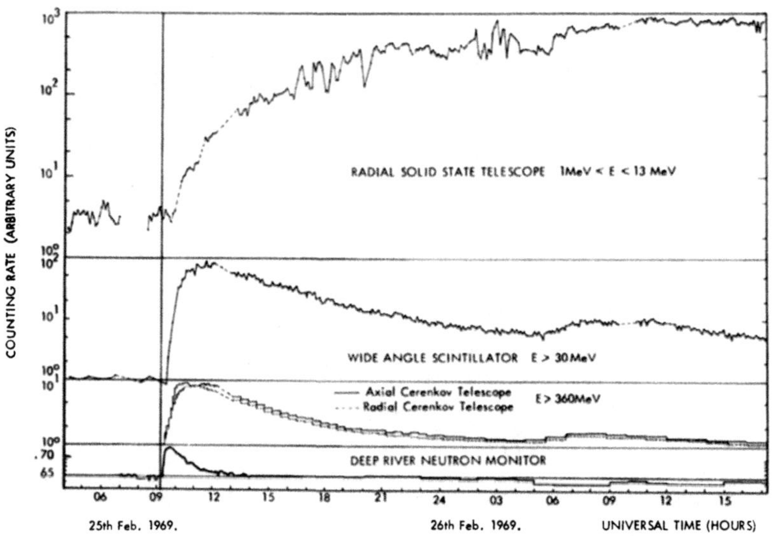


Fig. 8. Solar proton event of February 25, 1969 recorded by detectors on board HEOS 1. The Deep River neutron data are shown for comparison.

Figure 9 shows low energy proton data for the solar flare of May 28, 1967 (Lanzerotti, 1969). This flare was of importance 3B and reached its maximum development at 0545 in position W33 N28. The onset of the enhanced proton flux on an expanded time scale and for several energy bands is shown in Figure 10. These data show evidence for a delay in the arrival of the first particles which is greater than that corresponding to direct transmission along the average spiral field line. For instance, at 1.8 MeV the estimated travel time is 2h 50m for the spiral, whereas, the observed time delay between flare maximum and first arrival of these protons is $\approx$3h 50 m. Taken at their face values these figures would allow a storage time in the corona up to 60 min.

To summarize, then, there is evidence for delay in arrival of solar protons over and above that which would be expected on the basis of direct transmission along the idealised spiral of the average quiet time field. This delay could on occasion be as much as one to two hours, but is sometimes certainly very much less. It must be emphasised, however, that this time could have been spent in interplanetary space anywhere between the sun and earth and it is by no means established whether the particles spend the extra time on the close field lines of the lower corona or are merely confined behind an outward moving shock in the solar wind.

### 4.2. Intensity of Solar Particles as a Function of Direction

Storage in the corona cannot account for the long-lived background flux of protons which lasts for almost ten days. It is unlikely that protons in this energy range (0.6–13 MeV) can have been stored in the coronal fields for this time, since collision loss would set a limit lower than this. Neither could they be stored in the solar wind between the sun and earth for this period, since the solar wind would have carried them beyond the orbit of earth in this time. There remain three possibilities, that there are frequent new injections of freshly accelerated particles, that there is long term storage

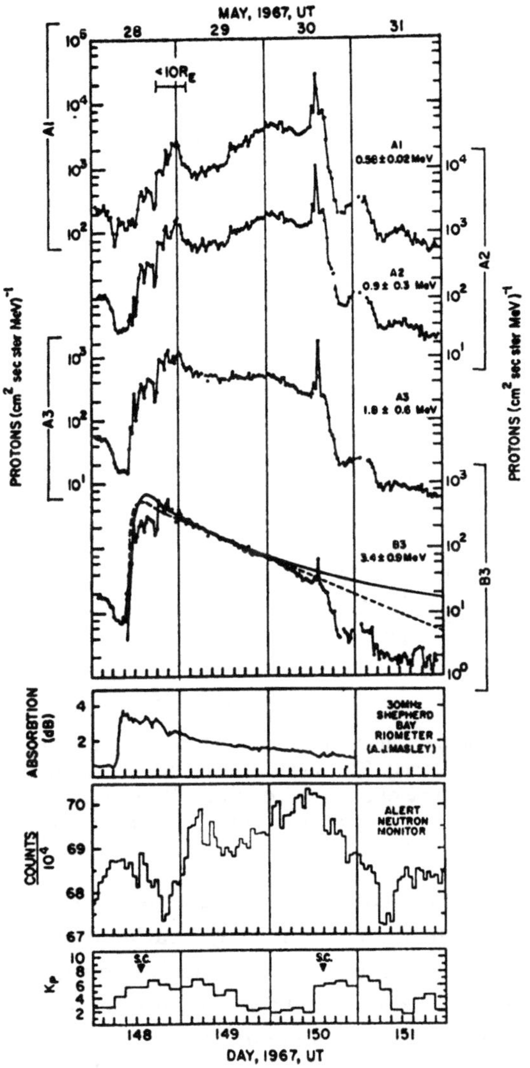


Fig. 9. Solar proton data averaged over 30-min intervals for period 28 to 31 May 1967 inclusive. The solid line and the dashed line through the B3 channel data are derived from theoretical models (after Lanzerotti 1969).

in a region containing the earth and which extends far beyond the orbit of earth, or both. The degree of anisotropy of these particles provides, in principle, a means of distinguishing between the alternatives.

If there is frequent acceleration of protons in the neighbourhood of the active solar region and the subsequent motion through the interplanetary magnetic field is

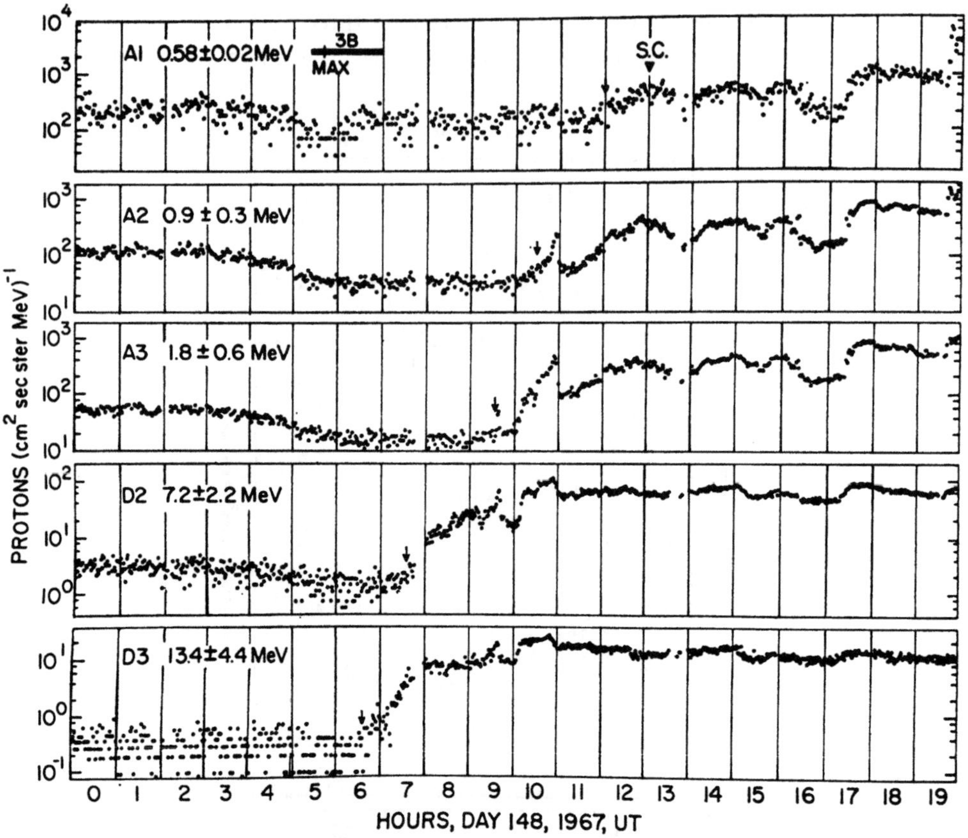


Fig. 10. Fast time resolution data showing the onset of the event on May 28, 1967. Each data point corresponds to one 9.28-sec interval. The small arrows indicate the time at which flare-enhanced fluxes were first observed. The time of observation of the 3B flare is noted at the top, the vertical line indicating maximum flare intensity at 0545 UT (after Lanzerotti, 1969).

adiabatic, as would be expected at the low energies, they should be rather well collimated along the field directions in the neighbourhood of the earth's orbit. Under these circumstances we should expect the particle flux to show a marked anisotropy. There have been a number of directional flux measurements of solar protons (see, for example, Balogh and Hynds, 1970; Fan *et al.*, 1968; McCracken *et al.*, 1968; Vernov *et al.*, 1969) and the general conclusion is that highly anisotropic flux distributions aligned with the field are only present for protons of a few MeV in the early stages of the flux enhancement and that after a period of a day or so the anisotropy reduces to something like 10% with maximum flux directed nearly radially outwards from the sun. The amplitude and direction of this latter type of anisotropy is consistent with containment of the particles in an extended trapping region where the distant trapping boundary moves outwards at the solar wind velocity but is not consistent with a source of particles located at the sun. This would indicate that at late times during impulsive events there is no large contribution to the flux from particles generated in a continuous acceleration process or from slow leakage of stored particles from the corona into interplanetary space.

## 5. Conclusions

(1) The complicated nature of charged particle propagation in interplanetary space makes it extremely difficult to deduce anything about particle diffusion in the corona from the study of the behaviour of energetic solar particles in the neighbourhood of the orbit of earth.

(2) There are no direct measurements of coronal magnetic fields but the fields calculated from measured photospheric fields and the assumption that the coronal fields are essentially current-free, allow estimates to be made of the relative importance of the various particle transport processes.

(3) The geometry of the calculated coronal magnetic fields and the matter density are such as to permit storage times of several days for protons in the energy range 10–100 MeV.

(4) The fields are of such a form as to permit rapid spreading along field lines of the particles in both latitude and longitude and this, coupled with the stochastic wandering of field lines, can readily account for the observed spread in solar longitude of the particles at 1 AU.

(5) The evidence on storage times derived from the onset of particle events is inconclusive, but intensity/time profiles and the directional intensity distributions that are observed would not be inconsistent with storage in the corona for periods of up to a day or so.

(6) The absence of marked anisotropies at late times during impulsive events is indicative of storage in interplanetary space rather than in the corona.

## References


Adams, N. and Braddick, H. J. J.: 1950, *Phil. Mag.* **41**, 505.

Altschuler, M. D. and Newkirk, G.: 1969, *Solar Phys.* **9**, 131.

Anderson, K. A.: 1969, *Solar Phys.* **6**, 111.

Axford, W. I.: 1965, *Planetary Space Sci.* **13**, 1301.

Balogh, A. and Hynds, R. J.: 1970, in *Intercorrelated Satellite Observations Related to Solar Events, Proc. 3rd ESRO/ESRIN Symposium, Noordwijk* (ed. by V. Manno and D. E. Page), Reidel Publ. Co., Dordrecht, p. 471.

Berry, E. B. and Hess, V. F.: 1942, *Terr. Magn. Atm. Electr.* **47**, 251.

Bryant, D. A., Cline, T. L., Desai, U. D.: 1962, *Phys. Rev. Letters* **14**, 481.

Cline, T. L. and McDonald, F. B.: 1968, *Solar Phys.* **5**, 507.

Ehmert, A.: 1948, *Z. Naturforsch.* **3a**, 264.

Elliot, H.: 1964, *Planetary Space Sci.* **12**, 657.
Elliot, H.: 1969, in *Solar Flares and Space Research* (ed. by C. de Jager and Z. Švestka), North-Holland Publ. Co., Amsterdam, p. 356.
Fan, C. Y., Pick, M., Pyle, R., Simpson, J. A., and Smith, D. R.: 1968, *J. Geophys. Res.* **73**, 1555.
Fisk, L. A. and Axford, W. I.: 1968, *J. Geophys. Res.*, **73**, 4396.
Holt, S. S. and Ramaty, R.: 1969, *Solar Phys.* **8**, 119.
Jokipii, J. R. and Parker, E. N.: 1969, *Astrophys. J.* **155**, 777.
Lange, I. and Forbush, S. E.: 1942, *Terr. Magn. Atm. Electr.* **47**, 331.
Lanzerotti, L. J.: 1969, *J. Geophys. Res.* **74**, 2851.
McCracken, K. G., Rao, U. R., and Ness, N. F.: 1968, *J. Geophys. Res.* **73**, 4159.
Newkirk, G.: 1967, *Ann. Rev. Astron. Astrophys.* **5**, 213.
Reid, G. C.: 1964, *J. Geophys. Res.* **69**, 2659.
Schatten, K. H.: 1968, *Nature* **220**, 1211.
Schatten, K. H.: 1969, *Nature* **223**, 652.
Schatten, K. H., Wilcox, J. M., and Ness, N. F.: 1969, *Solar Phys.* **6**, 442.
Tanaka, H. and Énomé, S.: 1970, *Nature* **225**, 435.
Vernov, S. N., Chudakov, A. E., Vakulov, P. V., Gorchakov, E. V., Kontor, N. N., Logachev, Yu. I., Lyubimov, G. P., Pereslegina, N. V., and Timofeev, G. A.: 1969, in *Proc. 11th Cosmic Ray Conference*, Budapest.
Wilcox, J. M., Severny, A., and Colburn, D. S.: 1969, *Nature* **224**, 353.
Zirin, H. and Severny, A.: 1961, *Observatory* **81**, 155.

# TIME VARIATIONS IN SOLAR ACTIVITY

HELEN W. DODSON and E. RUTH HEDEMAN
*McMath-Hulbert Observatory, The University of Michigan, U.S.A.*

**Abstract.** Long-term variations in the height of solar maxima and in the relative spottedness of the northern and southern solar hemispheres are considered. Mean sunspot numbers in the present solar cycle (number 20) are found to be similar to those in cycles with maxima in 1937, 1917, 1860, and 1769.

Time variations within the 11-year cycle are considered. It is pointed out that all solar latitudes and longitudes do not attain maximum levels of activity at the same time. Furthermore, the course of solar activity as measured by sunspot number and 2800 MHz flux is seldom smooth but appears to advance by series of pulses or episodes. The average time interval between onsets of successive Primary Fluctuations, as we have identified them in cycles 19 and 20, was 15 solar rotations. In general, Secondary Fluctuations occurred at intervals of 3 to 5 rotations. These statistical variations in sunspot number corresponded to recognizable episodes in specific solar activity and to maxima in plots of Major Flare and Active Region Indices.

Study of the declining branches of solar cycles 12 through 19 suggests that there will be 5 to 6 Primary Fluctuations of activity on the declining branch of cycle 20 in the years 1970.5–1975. At least half of these pulses can be expected to include significantly flare-rich centers of activity.

## 1. Long-Term Variations in the 11-Year Solar Cycle

### 1.1. Variations in Height of Sunspot Maximum

Observation of solar activity during the past five years with all of the techniques of ground-based and interplanetary equipment has been a challenging and wonderful experience for solar astronomers. Nevertheless, for those who remember the sun in the years around 1947 and 1958, there has been an element of disappointment. Although the sun has provided numerous instances of major activity, memory forces the realization that such events have been rare in comparison to the richness of solar activity in the two preceding decades.

In making the foregoing evaluation we have been experiencing directly the fact that individual solar cycles exhibit marked differences in the height of their respective maxima. Such differences are apparent from simple inspection of a plot of annual sunspot numbers. They represent long term, irregular variations in solar activity, and they are conveniently referred to as an apparent 80–100 year period in solar activity. Moving 11-year means of the annual sunspot numbers demonstrate clearly the magnitude and the irregularity of this long term variation in solar activity. (See Figure 1.)

The existence of this type of variation has been noted and discussed in the literature for more than 100 years. As early as 1861, Wolf commented upon it. Investigations in more recent years include those of Gleissberg (1945), Waldmeier (1957) (1966), Chernosky and Hagan (1958), and Vitinskii (1962). Kopecký in 1960 points out that this long term variation in sunspot number seems to stem from a cyclical change in the duration of sunspots rather than from a variation in the frequency of spot formation.

*Dyer (ed.), Solar Terrestrial Physics / 1970: Part I,* 151–172. *All Rights Reserved.*

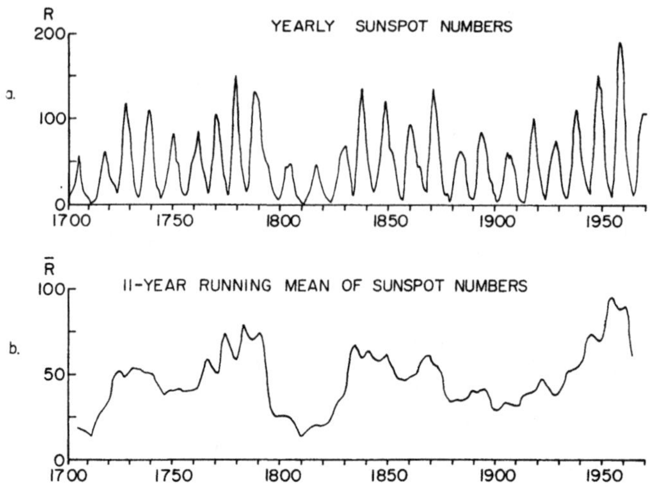


Fig. 1. Zurich sunspot numbers, 1700–1969. a. Annual sunspot numbers. b. 11-year running mean of annual sunspot numbers.

Superposed on the 80–100 year fluctuation in the level of solar activity is a somewhat systematic alternation of high and low maxima for successive 11-year cycles which suggests a possible relationship to the 22-year cycle in spot magnetic fields. The maxima of even numbered cycles (in the Zurich system) have tended to be lower than those of the odd numbered cycles, but there have been departures from this pattern.

### 1.2. Time variations in relative spottedness of northern and southern solar hemispheres

More or less associated with the 80–100 year variations in the level of solar activity are long term differences in the relative spottedness of the northern and southern hemispheres of the sun. This peculiarity of the sun has been known for a long time and has been studied by many people including Maunder (1904), Newton and Milsom (1955), and Waldmeier (1966). At times the difference between the spottedness of the northern and southern hemispheres of the sun has been extreme. Apparently there is no record of a northern sunspot in the 32-year interval, 1672 to 1704, and Cassini and Maraldi in 1705 are said to have commented specifically that they had never before seen a spot in the northern hemisphere (Maunder, 1890).

According to the studies of Newton and Milsom (1955) and Waldmeier (1957, 1966) there often is a phase, as well as an amplitude difference, in the spottedness of the northern and southern hemispheres. The former authors have shown that from 1833 to 1954, maximum spottedness frequently occurred at slightly different times in the northern and southern hemispheres, and that cycles with high maxima tended to be those in which the southern hemisphere was more spotted at the beginning of the cycle, and the northern hemisphere, at the end.

### 1.3. Comparison of cycle 20 with other cycles

The present solar cycle, Number 20, is illustrative of all of these various aspects of long term, time variations in solar activity. It is on the descending branch of the apparent 80–100 year period and is an even numbered cycle, circumstances that would presage lower activity than in cycle 19. According to sunspot numbers, cycle 20, with maximum smoothed monthly mean of 110.8 in November 1968, appears to resemble cycle 17 (max. 1937), cycle 15 (max. 1917), cycle 10 (1860) and cycle 2 (1769). (See Figure 1.)

A very strong north-south asymmetry has existed in cycle 20 from its beginning. Activity for this cycle started in 1963 in the northern hemisphere, and the north remained dominant at least through 1969. (See Figure 2.) According to the studies

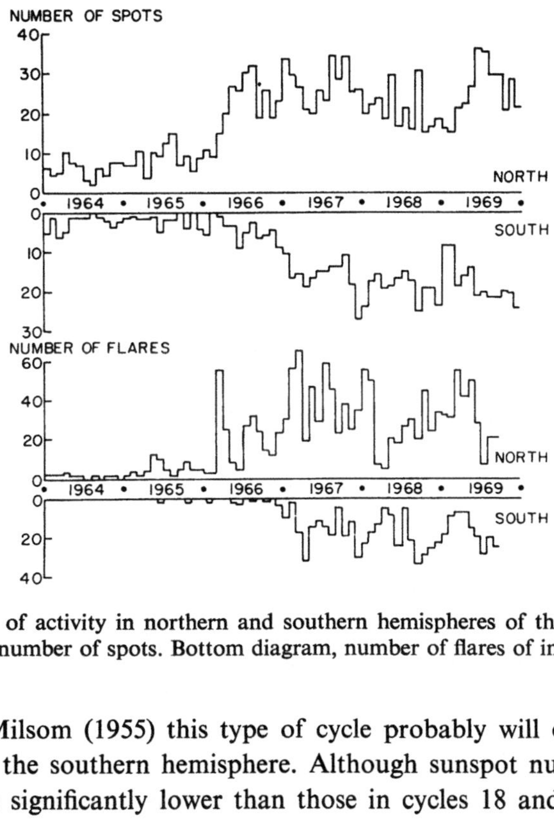


Fig. 2. Comparison of activity in northern and southern hemispheres of the sun, solar cycle 20. Top diagram, number of spots. Bottom diagram, number of flares of importance $\geqslant 1$.

of Newton and Milsom (1955) this type of cycle probably will close with activity predominantly in the southern hemisphere. Although sunspot numbers in cycle 20 are proving to be significantly lower than those in cycles 18 and 19, they are still above average values for the totality of cycles 1–19.

### 1.4. Extension of studies of solar activity to the distant past

It should be emphasized that detailed sunspot observations cover too short a time interval to permit more than recognition of the fact that major variations in intervals of 80 to 100 years have occurred during the past three centuries in the type and level of solar spottedness in successive 11-year cycles. Efforts have been made by numerous investigators including Schove (1955) and Bray (1968) to extend information about solar activity to the distant past through studies of ancient reports of sunspots, and of auroral and other terrestrial phenomena.

A recent analysis by Zhukov and Muzalevskii (1969) of the data assembled by Schove gives interesting results. The study indicates that Gleissberg's (1952) suggestion of a cycle of $\approx$1000 years is supported, a cycle with a period of $\approx$200 years is detected, but statistical evidence for the apparently observed 80–100 year variation is not found in the analysis. It is clear that the difficulties and uncertainties in all long range predictions of solar activity remain very great.

## 2. Time Variations Within the 11-Year Cycle

### 2.1. General comparison of ascending and descending branches of the solar cycle

The 11-year solar cycle is usually divided into times of generally increasing sunspot numbers and those with decreasing levels of solar activity. Numerous relationships have been determined between the duration of these ascending and descending branches of the cycle and the height of the respective maxima. Waldmeier (1935, 1966) has shown that for most, but not all, of the observed cycles, the plot of sunspot number versus time is asymmetric, with the interval from minimum to maximum being shorter than that from maximum to minimum. The greater the height of sunspot maximum, the shorter the duration of the ascending branch and the stronger is this type of asymmetry for the respective cycles. Waldmeier (1966) also comments that at the beginning of a cycle the rate of increase in sunspot number is steadily augmented up to a critical point where it suddenly drops, while activity continues to mount further to an ultimate maximum. On the contrary, the declining branch has a more uniform rate of descent.

Although the sun as a whole can be said to reach maximum activity at a particular time, all parts of the sun do not necessarily attain maximum at this time. It already has been pointed out that the northern and southern hemispheres frequently do not have simultaneous maxima. In a later section of this paper it will be shown that, in the present solar cycle, different longitudinal halves of the sun apparently reached maximum levels of activity at different times. In 1963, Gnevyshev, from studies of the corona, prominences, and sunspots, emphasized that the time of maximum is different for different latitudes. He considers the 11-year cycle to be the resultant of two processes, activity at high latitudes that attains a maximum early in the cycle and activity at low latitudes that reaches maximum one or two years later.

### 2.2. Fluctuations or pulses with durations of the order of 5 to 12 months

In addition to the foregoing well studied aspects of the 11-year cycle are other, poorly defined, shorter term fluctuations that influence greatly the level of solar activity at any specific time. The course of solar activity is seldom smooth but appears to advance by a series of pulses or episodes that endure or are separated in time by intervals of a rather small number of months. In an earlier study for the years 1963 September to 1967 May, we listed what seemed to us to be the times of marked onsets or critical changes in solar activity (Dodson and Hedeman, 1968) and commented that there appeared to be a roughly five-month cadence in the tabulation. Times of greater change occurred at intervals of approximately eleven months. These critical 'episodes' in the course of solar activity were selected subjectively on the basis of information relating to the development of major centers of activity, flare occurrence, the level of 10 cm radio emission, and the formation of large and complex spots.

Quantitative studies of strong fluctuations in relative sunspot numbers have been made by Vitinskii (1960a, 1960b, 1962). He reports that these fluctuations range in time from three months to a year, with an average duration of five months. Independently, Chernosky and Hagan (1958) selected 5 months as the most suitable time interval for the derivation of running means in sunspot numbers. Encouraged by these circumstances, we have attempted, for this paper, to explore further the tendency of the sun to show major changes in activity in a time pattern of a small number of months to a year.

## 3. Fluctuations or Pulses in Cycles 19 and 20

### 3.1. The principal fluctuations as exhibited by sunspot number and 2800 MHz flux

As a first step in efforts to study these short term fluctuations, we have brought together mean values of Zurich sunspot number and 2800 MHz flux for each Carrington rotation in solar cycles 19 and 20 (to date). (See Figures 3 and 4.) It is clear that the fluctuations shown by these indices of solar activity are generally similar. On the basis of these data, times of onsets of successive fluctuations have been determined according to our best judgment and information regarding the duration of the enhancements, and the interval between successive enhancements has been derived. (See Table I.)

In studying the fluctuations it seemed suitable to divide the cycles into three parts. The first interval covered the time from minimum through a rapid rise to an initial high maximum in sunspot numbers. The second part included the time of relatively sustained high level activity, and the third phase was the lower level, declining portion of the cycle. These divisions are indicated in Figures 3 and 4.

In addition, we once again saw in the data two types of fluctuations or pulses: Primary Fluctuations with durations of the order of a year, and Secondary Fluctuations which generally occurred with a cadence of three to five months. The Primary Fluctuations were clearly separated from each other by two or three rotations of marked diminution in the indices, or by 'no change', if on the ascending branch of the cycle. The Secondary Fluctuations were separated by smaller diminutions in activity which lasted only one or two rotations.

In cycles 19 and 20 (through February 1970), the Primary Enhancements, as we have identified them, ranged in duration from 8 to 17 rotations with a mean value

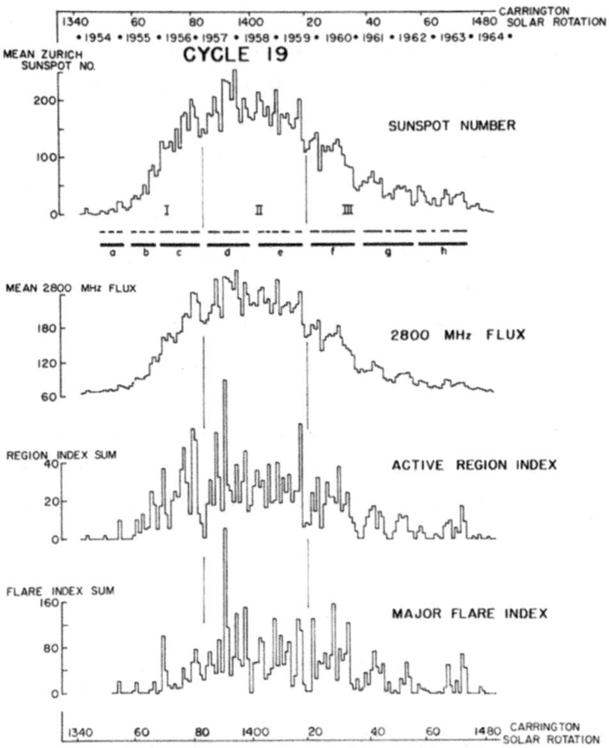


Fig. 3. Plot of various indices of solar activity for solar cycle 19. The data are recorded by Carrington rotations. The active region and major flare indices are described in the text. The broad dark lines, *a–h*, identify the primary fluctuations and the unlettered thin lines indicate the secondary fluctuations listed in Table I. The Roman numerals mark the three large divisions of the cycle.

of 13 rotations. The average time interval between onsets of successive Primary Fluctuations was 15 rotations. Secondary Fluctuations in our tabulation occurred at a mean interval of 4.6 rotations, with 70% of the cases falling within 3 to 5 rotations.

### 3.2. Comparison of Statistical Fluctuations with Specific Solar Activity

Do these statistical variations in sunspot number and 2800 MHz flux correspond to recognizable episodes in the course of solar activity? In an effort to gain a better understanding into what happens on the sun at the time of such fluctuations, we have prepared two new indices that reflect directly specific aspects of major solar activity.

#### 1. *Active Region Index*

The first of the new indices, the Active Region Index, refers to the occurrence of centers of activity that were above average in either flare production, ionizing radiation (SWF's), centimeter radiation, meter wavelength radiation, or size and complexity of spots. All centers of activity (i.e. all plages as identified by the McMath-Hulbert Observatory) in solar cycle 19, and in cycle 20 through January 1970, have been

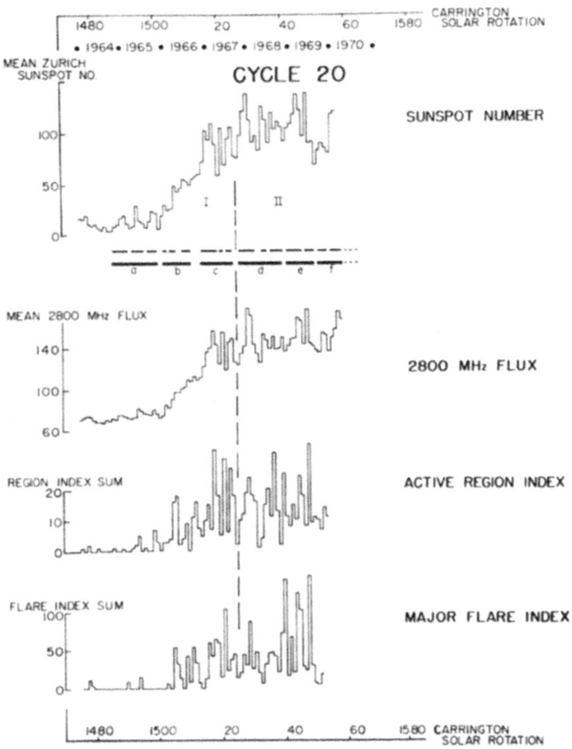


Fig. 4. Plot of various indices of solar activity for solar cycle 20, through 1969. The data are recorded by Carrington rotations. The active region and major flare indices are described in the text. The broad, dark lines, *a–f*, identify the primary fluctuations and the unlettered thin lines indicate the secondary fluctuations listed in Table 1. The Roman numerals mark the two large divisions of the cycle to date.

evaluated on the basis of these five parameters and have been listed in a catalogue. Importance evaluations ranging from 0 to 3, 4, or 5 were assigned to each region for each of the five constituents, and the Active Region Index is the sum of these components. Since the types of available radio frequency data changed from 1954 to 1970, we tried to modify the bases for our evaluations appropriately. Most centers of activity did not satisfy the criteria for inclusion in our 'Catalogue of Major Active Regions'. The catalogue lists only 298 regions in cycle 19 and 120 in cycle 20 through January 1970. The largest value of the index for any center of activity was 19. The Active Region Index was $\geqslant 10$ for only 59 regions in cycle 19, and 19 regions in the partially observed cycle 20. Summations of the Active Region Indices for the various regions that crossed the central meridian of the sun during each rotation have been plotted in Figures 3 and 4. No claims of excellence or suitability are made for the Active Region Index but it does provide a working tool for the recognition of the times of occurrence of major solar activity of the type described above. It is our understanding that these are the types of solar activity towards which much of the forecasting effort is currently directed.

TABLE I

Statistical fluctuations in rotational mean values of sunspot numbers and 2800 MHz flux
Solar cycles 19 and 20 (to April 1970)

| | | Carrington rotations | | | | | | |
|---|---|---|---|---|---|---|---|---|
| | | Primary fluctuations | | | Secondary fluctuations | | Dates | |
| | | Onset | End | | Onset | End | Onset | End |
| Cycle 19 Part I | Enhancement (*a*) | 1349 | 1356 | ($a_1$) | 1349 | 1350 | 1954 July 10 | Sept. 2 |
| | | | | ($a_2$) | 1352 | 1353 | 1954 Sept. 30 | Nov. 23 |
| | | | | ($a_3$) | 1355 | 1356 | 1954 Dec. 21 | 1955 Feb. 13 |
| | Rest or decline | 1357 | 1359 | | | | 1955 Feb. 14 | May 6 |
| | Enhancement (*b*) | 1360 | 1367 | ($b_1$) | 1360 | 1361 | 1955 May 7 | June 29 |
| | | | | ($b_2$) | 1363 | 1364 | 1955 July 27 | Sept. 19 |
| | | | | ($b_3$) | 1366 | 1367 | 1955 Oct. 17 | Dec. 10 |
| | Rest or decline | 1368 | 1369 | | | | 1955 Dec. 11 | 1956 Feb. 2 |
| | Enhancement (*c*) | 1370 | 1382 | ($c_1$) | 1370 | 1373 | 1956 Feb. 3 | May 22 |
| | | | | ($c_2$) | 1375 | | 1956 June 19 | July 15 |
| | | | | ($c_3$) | 1377 | 1378 | 1956 Aug. 12 | Oct. 5 |
| | | | | ($c_4$) | 1380 | 1382 | 1956 Nov. 2 | 1957 Jan. 22 |
| | Rest or decline | 1383 | 1385 | | | | 1957 Jan. 23 | Apr. 14 |
| Cycle 19 Part II | Enhancement (*d*) | 1386 | 1399 | ($d_1$) | 1386 | 1389 | 1957 Apr. 15 | Aug. 1 |
| | | | | ($d_2$) | 1391 | 1396 | 1957 Aug. 29 | 1958 Feb. 8 |
| | | | | ($d_3$) | 1398 | 1399 | 1958 Mar. 8 | May 1 |
| | Rest or decline | 1400 | 1402 | | | | 1958 May 2 | July 21 |
| | Enhancement (*e*) | 1403 | 1417 | ($e_1$) | 1403 | 1404 | 1958 July 22 | Sept. 14 |
| | | | | ($e_2$) | 1406 | | 1958 Oct. 12 | Nov. 7 |
| | | | | ($e_3$) | 1408 | 1409 | 1958 Dec. 6 | 1959 Jan. 28 |
| | | | | ($e_4$) | 1411 | 1413 | 1959 Feb. 26 | May 18 |
| | | | | ($e_5$) | 1416 | 1417 | 1959 July 12 | Sept. 3 |
| | Rest or decline | 1418 | 1420 | | | | 1959 Sept. 4 | Nov. 24 |
| | Enhancement (*f*) | 1421 | 1435 | ($f_1$) | 1421 | 1423 | 1959 Nov. 25 | 1960 Feb. 14 |
| | | | | ($f_2$) | 1425 | 1435 | 1960 Mar. 14 | 1961 Jan. 6 |
| | Decline or rest | 1436 | 1438 | | | | 1961 Jan. 7 | Mar. 29 |

| | | | | | | | | |
|---|---|---|---|---|---|---|---|---|
| Cycle 19 Part III | Enhancement ($g$) | 1439 | 1455 | ($g_1$) | 1439 | 1445 | 1961 Mar. 30 | Oct. 6 |
| | | | | ($g_2$) | 1447 | 1448 | 1961 Nov. 3 | Dec. 27 |
| | | | | ($g_3$) | 1450 | 1455 | 1962 Jan. 24 | 1962 July 6 |
| | Decline or rest | 1456 | 1457 | | | | 1962 July 7 | Aug. 29 |
| | Enhancement ($h$) | 1458 | 1474 | ($h_1$) | 1458 | 1460 | 1962 Aug. 30 | Nov. 19 |
| | | | | ($h_2$) | 1463 | | 1963 Jan. 14 | 1963 Feb. 9 |
| | | | | ($h_3$) | 1466 | 1468 | 1963 Apr. 6 | June 26 |
| | | | | ($h_4$) | 1470 | 1474 | 1963 July 24 | Dec. 6 |
| | Decline or rest | 1475 | – | | | | | |
| Cycle 20 Part I | Enhancement ($a$) | 1487 | 1500 | ($a_1$) | 1487 | 1491 | 1964 Oct. 29 | 1965 Mar. 14 |
| | | | | ($a_2$) | 1493 | 1496 | 1965 Apr. 11 | July 28 |
| | | | | ($a_3$) | 1498 | 1500 | 1965 Aug. 25 | Nov. 14 |
| | Decline or rest | 1501 | | | | | 1965 Nov. 15 | Dec. 12 |
| | Enhancement ($b$) | 1502 | 1511 | ($b_1$) | 1502 | 1503 | 1965 Dec. 13 | 1966 Feb. 4 |
| | | | | ($b_2$) | 1505 | 1506 | 1966 Mar. 5 | Apr. 27 |
| | | | | ($b_3$) | 1509 | 1511 | 1966 June 22 | Sept. 10 |
| | Decline or rest | 1512 | 1514 | | | | 1966 Sept. 11 | Dec. 1 |
| | Enhancement ($c$) | 1515 | 1524 | ($c_1$) | 1515 | 1519 | 1966 Dec. 2 | 1967 Apr. 17 |
| | | | | ($c_2$) | 1521 | | 1967 May 15 | June 10 |
| | | | | ($c_3$) | 1523 | 1524 | 1967 July 8 | Aug. 31 |
| | Decline or rest | 1525 | 1526 | | | | 1967 Sept. 1 | Oct. 24 |
| Cycle 20 Part II | Enhancement ($d$) | 1527 | 1540 | ($d_1$) | 1527 | 1531 | 1967 Oct. 25 | 1968 Mar. 9 |
| | | | | ($d_2$) | 1534 | 1535 | 1968 May 4 | June 26 |
| | | | | ($d_3$) | 1537 | 1540 | 1968 July 24 | Nov. 9 |
| | Decline or rest | 1541 | | | | | 1968 Nov. 10 | Dec. 7 |
| | Enhancement ($e$) | 1542 | 1550 | ($e_1$) | 1542 | 1546 | 1968 Dec. 8 | 1969 Apr. 22 |
| | | | | ($e_2$) | 1548 | 1550 | 1969 May 20 | Aug. 9 |
| | Decline or rest | 1551 | | | | | | |
| | Enhancement ($f$) | 1552 | $>1559$ | ($f_1$) | 1552 | 1554 | 1969 Sept. 6 | 1969 Nov. 26 |
| | | | | ($f_2$) | 1556 | $>1559$ | 1969 Dec. 24 | $>$1970 Apr. 12 |

## 2. *Major Flare Index*

The second of the solar activity indices that we have used to supplement the sunspot and 2800 MHz data is a Major Flare Index. This index has been derived for all flares for which any one of the following circumstances apply: optical importance $\geqslant 3$; SWF importance $\geqslant 3$; 10 cm radio event with flux $\geqslant 500 \times 10^{-22}$ Wm$^{-2}$ (c/sec)$^{-1}$; Type II or Type IV radio frequency emission; occurrence of Polar Cap Absorption or ground level cosmic ray increase. For each such flare, importance evaluations were derived for five parameters related to the optical, ionospheric, and radio frequency aspects of the flare and were added together to form the Major Flare Index.

From January 1955 to September 1969 there were 982 flares that qualified for calculation of the Major Flare Index. The largest value of the index for any single flare was 17. For 164 flares in cycle 19 and for 46 flares in the partially observed cycle 20, the index was $\geqslant 10$. Summations of the Major Flare Index have been made by region, and by the Carrington rotation in which the respective regions crossed the central meridian of the sun. The resultant values are plotted in Figures 3 and 4.

## 3. *Description of Specific Activity at Times of Certain Statistical Fluctuations*

Comparison of the statistical fluctuations in sunspot and 2800 MHz data with the new solar activity indices and with detailed primary reports of solar events indicates that, in general, the statistical variations derived from spot and radio-frequency data, and entered in Table I, correspond to recognizable episodes in the course of solar activity. The following examples will serve to illustrate the types of activity on the sun that accompanied these statistical fluctuations:

*Cycle 20 Part I: Fluctuation $b_2$; rotations 1505–06; 1966 March 5 – April 27*

This interval, to be referred to as 20-$b_2$, is the second enhancement or pulse in the Primary Fluctuation that is considered to have started in mid-December 1965 and to have lasted through November 1966. Like other Secondary Fluctuations on the ascending branch of the cycle, 20-$b_2$ was difficult to identify unambiguously. Increases in the 2800 MHz flux in rotation 1505 and in the spot number in 1506 provided the bases for the listing of this fluctuation. On the other hand, the Active Region and Major Flare Indices mark it as a conspicuous enhancement with a level of flare activity comparable to that in months near solar maximum. Specifically, the interval included the development of two major centers of activity on essentially opposite sides of the sun. The first region, N20° CMP March 22, with a $\gamma$-type spot and 127 flares of importance $\geqslant 1$ was associated with 31 SWF's and with energetic particle emission. The second major center of activity, N27° CMP April 3, produced 64 flares of importance $\geqslant 1$, and 35 SWF's. The interval also marked the outbreak of numerous spots and plages which together formed the encircling northern member of the Zone Royale.

This fluctuation, 20-$b_2$, is separated from its successor, 20-$b_3$, by two rotations of 'rest' or lessening of activity, primarily in May and June 1966. Fluctuation 20-$b_3$

was dominated by the activity associated with, and subsequent to, the proton and cosmic ray flare of 1966 July 7. It should be remembered that the fluctuations and enhancements here identified depend only on the increases, maxima, and minima in sunspot number and 2800 MHz flux for the sun as a whole, and do not refer to the detailed study of the exact time of onset and duration of individual centers of activity.

*Cycle 20 Part I: Fluctuation $c_1$; rotations 1515–19; 1966 December 2 – 1967 April 17*

This fluctuation, 20-$c_1$, is conspicuous and well defined in both sunspot and 2800 MHz data. It marks the occurrence of a new high level of activity in these parameters for solar cycle 20. It is a level of activity roughly comparable to that exhibited by solar cycle 20 during the next three years. It is in this pulse or fluctuation that major activity in the southern hemisphere developed for the first time. (See Figure 2.) The first southern region in cycle 20 to qualify for inclusion in our catalogue of Major Centers of Activity, crossed the central meridian on 1966 December 11 at S23°. It was followed in the next rotation (1516) by the development of two more major centers of activity in the south, S22° CMP January 2 and S23° CMP January 16. Three additional southern regions with major activity traversed the disk during fluctuation 20-$c_1$. It seems correct to say that during this fluctuation the southern member of the Zone Royale developed.

Although activity in the south was the new characteristic of fluctuation 20-$c_1$, the greatest source of activity in the interval was in northern latitudes. The two transits of the region at N22° longitude 295°, with central meridian passages on February 27 and March 27, respectively, produced successively 97 and 75 flares of importance $\gg 1$.

It seems worthy of note that the unusual ground level enhancement of cosmic rays on 1967 January 28, presumably from a region on the invisible hemisphere (Dodson and Hedeman, 1969), occurred during this fluctuation, which marked the first time in cycle 20 when the sun exhibited major activity in both the northern and southern hemispheres. There were four additional instances of Polar Cap Absorption within the fluctuation.

Fluctuation 20-$c_1$, represented a consistently high level of activity for 5 rotations. It was then followed by an interval of more than a rotation in which the daily sunspot number did not rise above 100 or the daily 2800 MHz flux above 140 even though calcium spectroheliograms showed the continued presence of northern and southern bands of plages. This protracted interval of relative calm was broken by the development of the well known particle producing center at N26° CMP 1967 May 25 in the next fluctuation.

*Cycle 20 Part II: Fluctuation $d_3$; rotations 1537–1540; 1968 July 24 – November 9*

The Secondary Fluctuation, 20-$d_3$, occurs during the years of maximum activity and comes at the end of a Primary Fluctuation that began ≈8 months earlier. 20-$d_3$ is a well defined fluctuation in sunspot numbers, but according to these data, its maximum was lower than those of the two preceding Secondary Fluctuations, 20-$d_1$, and 20-$d_2$. The principal activity in fluctuation 20-$d_3$ takes place during the last rotation, 1540, which included the central meridian passage of two great centers of activity N17° CMP October 23 and S15° CMP October 28. These regions were associated with the many energetic phenomena described in the report UAG-8, World Data Center A (ESSA), entitled 'Data on Solar-Geophysical Activity, October 24 – November 6, 1968'. In rotation 1540, the Active Region Index was high and the Major Flare Index reached the highest value up to that time in cycle 20. On the other hand, in this rotation, both mean and daily values of sunspot number and 2800 MHz flux failed to attain the high levels that might have been expected to accompany the occurrence of regions of such unusual activity.

Fluctuation 20-$d_3$ was followed by a minimum in both sunspot number and 2800 MHz flux in rotation 1541. In identifying the fluctuations for Table I, this rotation was selected as an interval of relative quiet between two Primary Fluctuations, 20-*d* and 20-*e*. It is interesting that in spite of the relatively low values of the various overall solar parameters for rotation 1541, an energetic particle event took place within it on 1968 November 18 in time association with a major flare with Index of 7. The event serves to emphasize the point of view that energetic particle events often depend on circumstances other than unusually high levels of the usual measures of solar activity.

*Cycle 20 Part II: Fluctuation $e_1$; rotations 1542–1546; 1968 December 8 – 1969 April 22*

This Secondary Fluctuation, also during the years of maximum activity, lasted for 5 rotations and reached its highest levels gradually. The mean sunspot number for rotation 1545 (1969 February 28 – March 26) was 138, one of the three highest values recorded up to this time (1970 March) in cycle 20. In addition, at the start of rotation 1546, 27-day running means of the daily 2800 MHz flux rose to maximum values for cycle 20 (to 1970 March).

Fluctuation 20-$e_1$ was an interval characterized by activity in two large zones in approximately opposite hemispheres of the sun. The lesser zone of activity was centered at $\approx 260°$ Carrington longitude, and the greater zone at $\approx 75°$. This greater zone, at days 22–23 in the Carrington rotation, was associated with proton events in 1969 January, February, March, and April, with the ground-level cosmic ray increase of 1969 February 25, and with the long interval of high proton flux observed by Explorer 24 starting 1969 April 10.

Fluctuation 20-$e_1$ represented an interval of high and sustained activity lasting for 5 rotations. It was followed in rotation 1547 by a conspicuous diminution in all of the solar parameters here considered. This interval of relative calm was then broken by a resurgence of activity in 1969 June in the active longitude zone ($\approx 80°$) referred to above. The time pattern of these events resembles that of fluctuation 20-*c*, described earlier.

*Cycle 19 Part III: Fluctuation $f_2$; rotations 1425–1435; 1960 March 14 – 1961 January 6*

This fluctuation of long duration (11 rotations) is well defined as an entity in both sunspot and 2800 MHz data and does not invite separation into shorter intervals. By the time of this fluctuation, 2800 MHz flux above quiet sun levels and mean sunspot numbers had dropped to about half their values at the time of maximum in the cycle. On the other hand, the Active Region and Major Flare Indices for the fluctuation were still almost as high as during years associated with greatest sunspot activity. (See Figure 3.)

Fluctuation 19-$f_2$ begins with the transit across the solar disk in rotation 1425 of one of the great centers of activity of solar cycle 19. This region at N12° longitude 135° traversed the disk three times as a flare-rich region with a $\gamma$-type spot, with successive crossings of the central meridian on 1960 March 31, April 27, and May 24. During its March transit of the disk, it was the site of four proton producing flares, and in its next transit, a ground-level cosmic ray enhancement as well as an additional proton flare were associated with the region. Another important center of activity developed during the early part of fluctuation 19-$f_2$. This region with longitude 8° was at the unexpectedly high latitude of N29°. It contained a $\gamma$-type spot with maximum area 2331 millionths of the solar hemisphere. This was the largest spot of 1960 and the last spot of cycle 19 to have area $\geqslant$ 2000 millionths. For two of its transits across the solar disk (CMP 1960 May 7 and June 4) it was the site of many flares, one of which caused PCA on May 13.

High levels of flare production continued through 19-$f_2$ with the transit of the following regions, each of which had more than 30 flares of importance $\geqslant 1$: N20° CMP June 25, N10° CMP July 30, N20° CMP August 13, N17° CMP August 31, S15° CMP September 25. Activity in this fluctuation culminated with the transit on 1960 November 12 (rotation 1433) of the great flare and energetic particle producing region at N24° longitude 30°. This region is well known as one of the greatest centers of solar activity of cycle 19. It is noteworthy that neither the mean nor the daily values of sunspot number and 2800 MHz flux during rotation 1433 showed increases that would suggest the occurrence in November 1960 of this unusually great source of specific solar activity. On the other hand, the Active Region Index for the center was 17, a value matched by only three other regions in solar cycle 19. The Major Flare Index for rotation 1433 also was comparable to values of this index for most of the times of high flare activity during the years of sunspot maximum in cycle 19.

After the close of events associated with the 1960 November 12 region, the level of activity dropped quickly in rotations 1436–1438 (1961 January 7 – March 29) to new low levels of activity for the declining branch of cycle 19. In summary, fluctuation 19-$f_2$, about two years after sunspot maximum, both began and ended with the occurrence of centers of activity that were as abundant in flares and energetic particle events as any in the entire cycle.

*Cycle 19 Part III: Fluctuation $g_1$; rotations 1439–1445; 1961 March 30 – October 6*

Like other fluctuations on the declining branch of the cycle, 19-$g_1$ is well defined by both sunspot number and 2800 MHz flux. It also can be recognized as an entity in plots of the Major Flare and Active Region Indices. (See Figure 3.) Since the preceding fluctuation (19-$f_2$ described above) the average sunspot number and the 2800 MHz flux above the quiet sun, had diminished by another factor of two. All regions with significant activity were located on one side of the sun. During the fluctuation, 15 regions with Active Region Index $\geqslant 1$ traversed the solar disk but not one of them was located within the 160° of longitude extending from 90° to 250°.

Three flare-rich regions occurred during the fluctuation S10° CMP July 14, N12° CMP August 17, and N15° CMP September 4, with 79, 41, and 69 flares of importance $\geqslant 1$, respectively. The first of these regions was associated with multiple proton enhancements and two ground level increases in cosmic rays. PCA was recorded also with a flare in the September 4 region and with a less flare-rich region with CMP on October 1.

Activity during fluctuation 19-$g_1$ illustrates clearly the fact that a small number of significantly active regions with major flares and energetic particle events occur within statistical fluctuations on the descending branch of the cycle, even though the general level of solar activity as shown by spot number and 2800 MHz flux may be relatively low.

In summary, the foregoing descriptions of specific solar activity at the time of statistical fluctuations or pulses in sunspot and 2800 MHz data indicate that these fluctuations generally identify intervals in which enhancements of major flare and particle activity have taken place. It is noteworthy, however, that the statistical fluctuations fail to provide dependable information about the magnitude or level of the specific activity. The times of great flare-rich and/or particle producing centers occurred variously at the beginning, middle, or near the end of individual fluctuations.

The time pattern of the statistical fluctuations in sunspot and 2800 MHz data apparently gives a first approximation to time variations in the occurrence of major, flaring centers of activity. According to our studies, these successive pulses or episodes in solar activity occur at intervals of four or five months to a year without any obvious regularity of pattern. On the other hand, if spot numbers and spot areas back to 1878 provide adequate guidance, such fluctuations are characteristic of the course of solar activity and are to be expected. Solar cycles apparently advance by a series of pulses or episodes rather than by an uninterrupted increase or decrease in the general level of activity.

## 4. Time Variations in the Longitude Distribution of Major Activity, Cycles 19 and 20

A review of the time variations in solar activity should not overlook a variation of a type rather different from those discussed in the foregoing sections of this paper. We refer to changes in the longitudes in which major activity apparently concentrates for long intervals of time. Figure 1 of a recent paper by Švestka and Simon (1969) provides an example of such a change in the location of the source of the maximum 2800 MHz flux as averaged over successive intervals of 12 rotations.

Consideration of the apparent occurrence of solar activity in preferred longitudes is not rightfully part of the topic assigned to this paper, but the time pattern of longitude distribution of activity in solar cycle 20 has been so sharp (through 1970 February) that it seems worthy of mention. Following in part the mode of presentation by Švestka and Simon, we have prepared average values of the daily 2800 MHz flux as a function of Carrington longitude for successive intervals, each 6 rotations long. In Figure 5, we plot these values versus time from rotation 1478 to 1555 (1964 February 27 to 1969 December 23). The diagram shows that from the beginning of the plot through rotation 1519, (1967 March 21) the maxima in the flux curves centered around day 10 in the Carrington rotation. In subsequent rotations the maxima cluster around day 23 the opposite side of the sun. The transition from one longitude to the other occurred in the interval covering rotations 1520 to 1525. Passage of the great flare-rich, proton-producing region with CMP on 1967 May 25 (rotation 1521) terminated the sequence of maxima centered around day 10. The first of the great regions with maximum flux near day 23 was the pair of centers of activity with CMP on 1967 July 28 and 29 respectively (rotation 1523). These latter regions were exceedingly flare-rich and abundant producers of ionizing radiation, but they were not associated with major proton events or geomagnetic storms. In later rotations regions with CMP near day 23 did produce energetic particles. For example, the proton and cosmic ray producing flare of 1969 February 25 was in a region near this longitude.

If the values of the daily 2800 MHz flux for the 10th and 23rd days of each Carring-

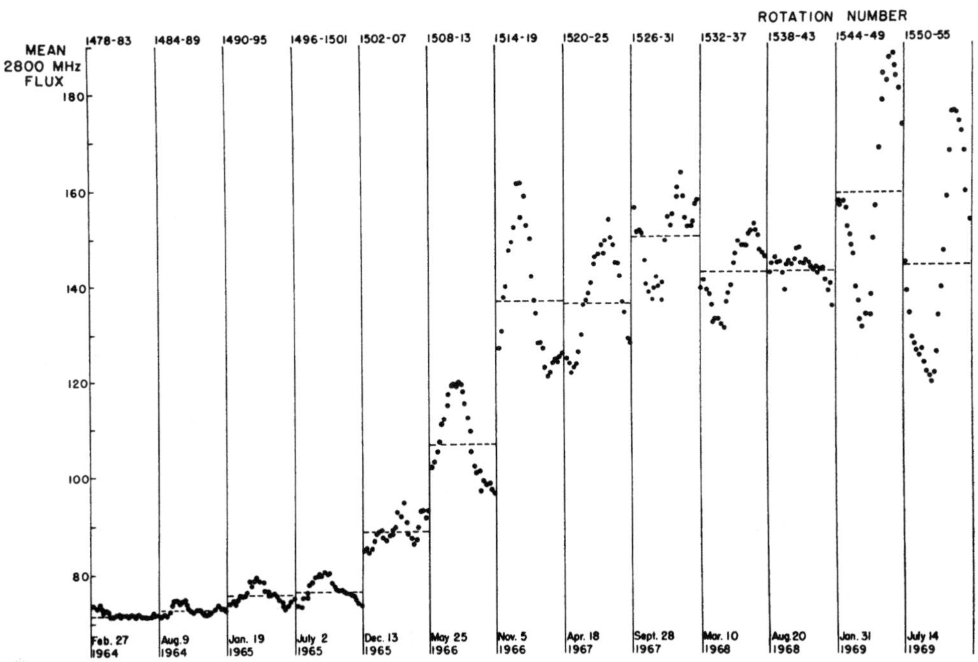


Fig. 5. Average values of 2800 MHz flux for each day of a Carrington rotation for successive intervals of six rotations, solar cycle 20 through 1969. Dashed lines indicate the mean value of the flux for the entire interval of six rotations. The dates indicate the start of the interval for which the averages have been derived.

ton rotation (or of 13 day means centered on these days) are plotted versus time, it becomes clear that the foregoing differences can be interpreted as a result of an earlier rise to maximum activity in longitudes clustered around day 10 than for those clustered around day 23. (See Figure 6.) What the end of solar cycle 20 will show, only time can tell.

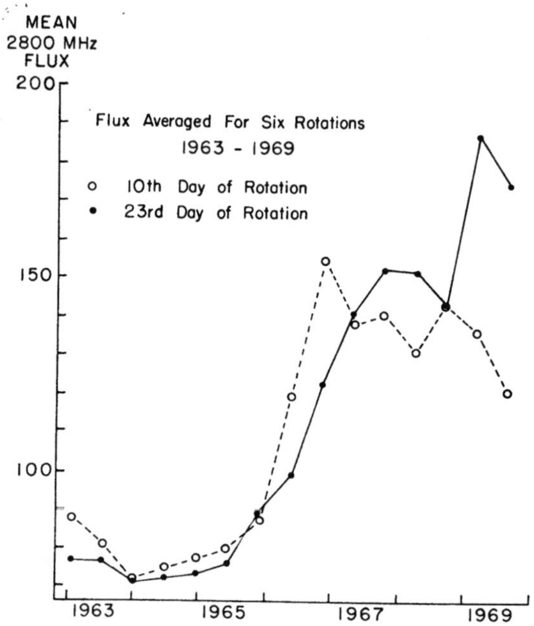


Fig. 6. Mean values of 2800 MHz flux for the 10th and 23rd days of Carrington rotations averaged for successive intervals of six rotations, 1963–1969.

For the higher and more active solar cycle 19, various indices of solar activity show evidence of preferred longitudes. According to the times of maxima in 2800 MHz flux and the times of central meridian passage of the centers of activity included in the Active Region Indices, the following patterns emerge. In the early part of the cycle, 1954 to mid 1957, activity was centered around days 11 to 18 of the Carrington rotation. For subsequent years up to 1962, the opposite hemisphere became dominant, and the favored longitude zone corresponded to days 22 to 27 and 1 to 7. In mid 1962, maximum activity began to center around day 10, and this pattern was maintained during the years of minimum and to mid 1967 on the ascending branch of cycle 20.

It must be emphasized that although long enduring, preferred zones of activity imply differences in the course of the solar cycle for different longitudes, description of these differences is dependent on the rate and type of rotation chosen to determine the longitude system. Carrington longitudes have been used in this paper. Furthermore, much of the discussion refers to differences in the level of activity in longitude zones on the sun and does not imply that significant activity occurred exclusively in the apparently 'favored' zones. Throughout the years studied, numerous interesting and flare-rich regions developed in the 'nonfavored' longitudes.

## 5. What Lies Ahead?

Do the foregoing efforts to study the time variations in solar activity, provide patterns sufficiently well established to allow one to comment with some assurance on what lies ahead in solar activity in the next five years? We think not, but in these days in which considerable emphasis is placed on the anticipation of solar activity, perhaps one should try to evaluate solar circumstances for this interval.

First, the level of activity can be expected to decline to minimum values sometime around 1975.5 or 1976.0. The exact time of minimum probably will depend more upon the time at which cycle 21 begins than upon the rate of decline of cycle 20.

Activity in the southern hemisphere of the sun may predominate. Additionally, if cycle 20 mimics cycle 19, another change in preferred longitude for major activity may occur in $\approx$1972.

A comparison of data for the declining branches of solar cycles 12 through 19 suggests that there will be 5 to 6 Primary Fluctuations of activity on the declining branch of cycle 20 in the years 1970.5–1975, and that at least half of these pulses will include the formation of significantly flare-rich centers of activity. (See Figures 7a, b.) In this diagram we have brought together for the last five years of cycles 12–18, sunspot areas as measured at Greenwich and information relating to large spots, major flares, and great geomagnetic storms as recorded in the volume 'Sunspot and Geomagnetic Storm Data, derived from Greenwich Observations, 1874–1954'. For cycle 19, the plot of sunspot area is replaced by 2800 MHz flux (Ottawa data) and the spot and flare information represent compilations of reports from Greenwich and Rome observatories and the Quarterly Bulletin on Solar Activity. During the declining phase of the cycle, the pulses or enhancements of activity are generally separated by at least 2 or 3 rotations of very low activity.

If past experience is a guide, a truly major center of activity may occur even as late in the cycle as $\approx$2.5 years prior to minimum. The corresponding time in cycle 20 probably will be 1973. A list of these great centers of activity late on the declining branch of past cycles is given in Table 2 and the events are shown graphically in Figure 8.

After 1973, the number of spots with maximum area $\geqslant$750 millionths of the hemisphere can be expected to diminish rapidly. In the last years of the cycle, such spots will occur only sporadically, one or two to each fluctuation or sustained enhancement in sunspot number. It should be noted, however, that the last five solar cycles produced a flare-rich region, or spots with maximum area $\geqslant$1000 millionths of the hemisphere, within a year prior to sunspot minimum. (See Figure 8).

Similar regularities seem to have occurred on the early ascending branches of the solar cycles. For 7 of the last 9 cycles, spots with maximum areas $\geqslant$ 1000 millionths of the hemisphere developed within approximately one year after minimum in sunspot

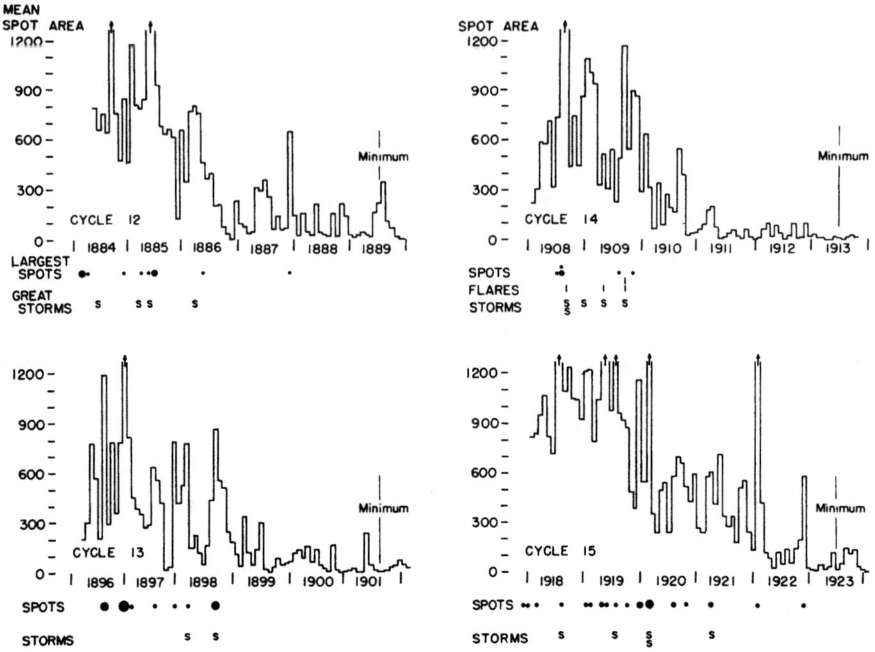


Fig. 7a.

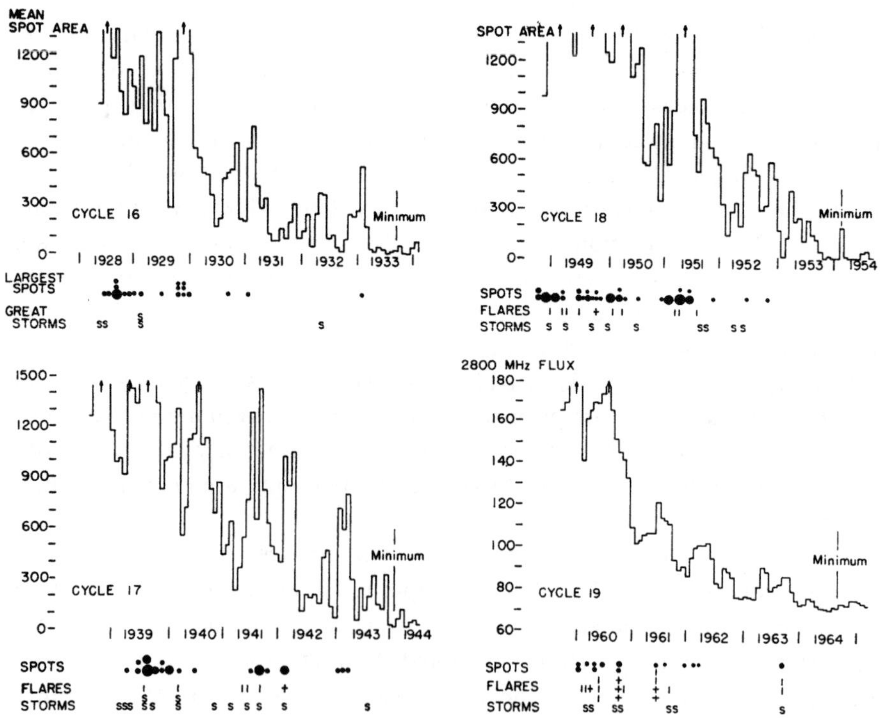


Fig. 7b.

Fig. 7. Solar activity on the declining branch of solar cycles 12–19. a. Cycles 12–15; b. Cycles 16–19. For cycles 12–18 mean spot areas (Greenwich data) are plotted for each Carrington rotation. For cycle 19, mean values of the 2800 MHz flux for each Carrington rotation are recorded. The largest spots and times of the greatest geomagnetic disturbance are shown for all cycles. For cycles 14, 17, 18, and 19 known major flares are indicated by a short vertical line. This line is crossed when a ground level enhancement of cosmic rays occurred.

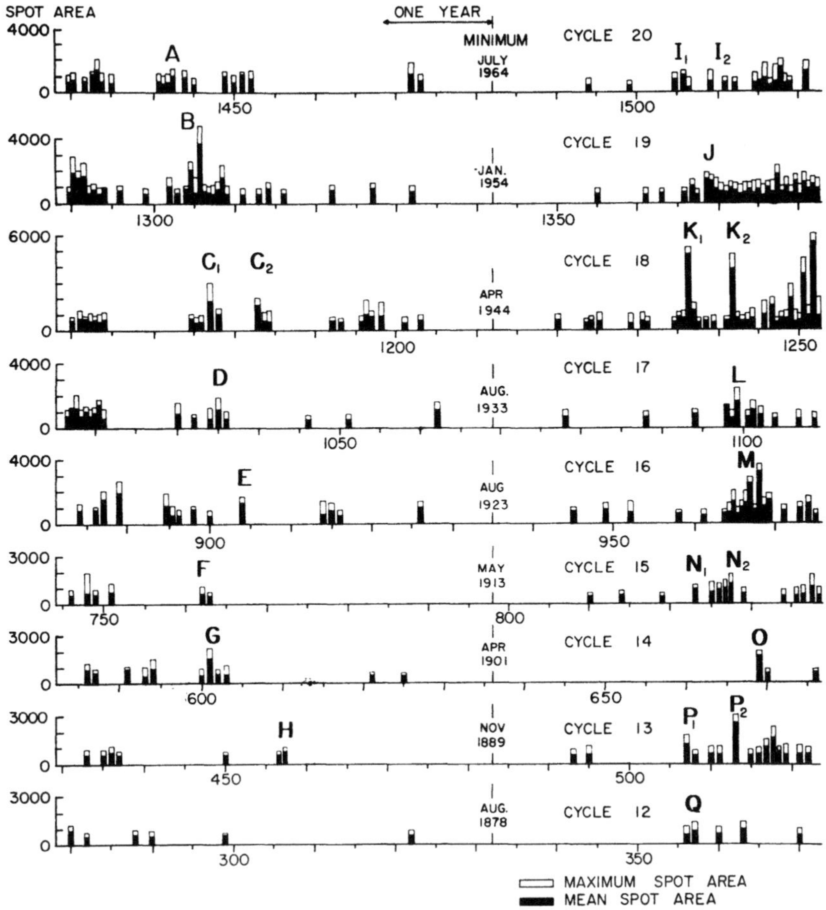


Fig. 8. Distribution by solar rotation of all spots with maximum area $\geqslant 750$ millionths of the hemisphere for $\approx 6$ years around solar minimum. Data taken from 'Sunspot and Geomagnetic-Storm Data, Greenwich Observations, 1874–1954' and from spot areas as reported in 'Greenwich Photoheliographic Results 1955–1961' and in the 'Monthly Bulletin of Solar Phenomena, Rome Observatory, 1962–1967'. Minimum as here given is based on consideration of lowest values of sunspot areas and is not the 'statistical' minimum of sunspot numbers. Letters *A* to *Q* refer to major solar activity $\approx 2.5$ years before and $\approx 2$ years after sunspot minimum. (See Table II.)

numbers. Within $\approx 2$ years after minimum for all cycles since 1878, truly great centers of activity had developed, with very large, complex spots and histories of major flares and particle emission when such information was being obtained. A list of these events is given in Table II and they are shown graphically in Figure 8.

Although the ability to anticipate the general course of solar activity is limited, the present study has found enough regularity in the data to warrant hope that in years ahead, greater insight in the problem may be attained. Further development of indices that reflect directly the specific types of solar activity for which the forecasting needs are most urgent, should add to the progress being made in this field.

TABLE II

Data concerning major solar activity approximately 2.5 years before and 2 years after sunspot minimum

| Cycle | Feature | Rotation | Date | $\Delta t$ to min.[a] | |
|---|---|---|---|---|---|
| | | | Descending branch of cycles 12–19 | | |
| 19 | A | 1442 | July 1961 | − 3 yr. | Transit of large spot in active, flaring region that produced numerous particle events and ground level cosmic rays |
| 18 | B | 1305–07 | April, May and June 1951 | − ≈ 2 yr. 8 mo. | Time of transit of three very large and complex spot groups. May transit most active and flare-rich. None of these 3 groups was associated with a great magnetic storm |
| 17 | $C_1$ | 1177 | Sept. 1941 | − 2 yr. 7 mo. | A large complex spot group. A very great flare on Sept. 17 was followed by an intense geomagnetic storm (one of the rare storms for which $K_p = 9$) with a great auroral display |
| 17 | $C_2$ | 1183 | Feb. 1942 | − 2 yr. 2 mo. | A big complex spot group with several notable eruptive flare events. A great flare on Feb. 28 was accompanied by cosmic rays and was followed by a great magnetic storm |
| 16 | D | 1035 | Feb. 1931 | − 2 yr. 6 mo. | A large spot group, that developed very rapidly on the disk |
| 15 | E | 904 | May 1921 | − 2 yr. 3 mo. | A large complex spot situated on the solar equator. When spot was near central meridian, a severe magnetic disturbance occurred, with a great auroral display |
| 14 | F | 762 | Oct. 1910 | − 2 yr. 7 mo. | Last spot with maximum area > 1000 millionths seen prior to minimum |
| 13 | G | 601 | Sept. 1898 | − 2 yr. 7 mo. | Transit of very large spot |
| 12 | H | 457 | Dec. 1887 | − 1 yr. 11 mo. | Last spot with area > 1000 millionths prior to minimum |
| | | | Ascending branch of cycles 12–20 | | |
| 20 | $I_1$ | 1505–06 | March and April 1966 | ≈ + 1 yr. 8 mo. | First very large spots and flare-rich regions since minimum |
| 20 | $I_2$ | 1509, 1511 | July and Aug. 1966 | ≈ + 2 yr. | Large complex spot groups in active and flare-rich regions, associated with energetic particles |
| 19 | J | 1369–70 | Jan. and Feb. 1956 | + 2 yr. | Very large spots. Great complex group of February associated with great cosmic ray event of February 23 |
| 18 | $K_1$ | 1236 | Feb. 1946 | + 1 yr. 10 mo. | Two of the largest spots ever observed, in very active and flare-rich regions, and associated with great magnetic disturbance. A very great flare, with cosmic rays, occurred on July 25 |
| 18 | $K_2$ | 1242 | July 1946 | + 2 yr. 3 mo. | |

*Table II (continued)*

| Cycle | Feature | Rotation | Date | $\Delta t$ to min.[a] | |
|---|---|---|---|---|---|
| 17 | L | 1099 | Dec. 1935 | +2 yr. 4 mo. | Very large and complex spot group, in active and flaring region |
| 16 | M | 965–67 | Nov. and Dec. 1925 and Jan. 1926 | ≈ +2 yr. 4 mo. | Very large and complex spot groups. Spot of January 1926 was largest ever recorded, until great spots of cycle 18 appeared in 1947 |
| 15 | $N_1$ | 823 | April 1915 | +1 yr. 11 mo. | First spot with area > 1000 millionths since minimum |
| | $N_2$ | 827 | July and August 1915 | ≈ +2 yr. 2 mo. | Very large spots |
| 14 | O | 669 | Oct. 1903 | +2 yr. 6 mo. | Very large spot, and first major spot since minimum |
| 13 | $P_1$ | 507 | Sept. 1891 | +1 yr. 10 mo. | Very large spots |
| | $P_2$ | 513 | Feb. 1892 | +2 yr. 3 mo. | |
| 12 | Q | 357 | June 1880 | +1 yr. 10 mo. | Large spot |

[a] The minimum here used is based on consideration of lowest values of sunspot areas and is not the 'statistical' minimum of sunspot numbers.

## Acknowledgements

The authors most gratefully acknowledge partial support in the preparation of this study from the following grants and contracts: NASA NGL-23-005-275, NSF GA-632, Nonr-1224(19), and N00014-67-A-0181-0001.

## References

Bray, J. R.: 1968, *Nature* **220**, 672.
Chernosky, E. J. and Hagan, M. P.: 1958, *J. Geophys. Res.* **63**, 775.
Dodson, H. W. and Hedeman, E. R.: 1968, in K. O. Kiepenheuer (ed.) 'Structure and Development of Solar Active Regions', *IAU Symp.* **35**, 56.
Dodson, H. W. and Hedeman, E. R.: 1969a, *Ann. IQSY* **4**, 3.
Dodson, H. W. and Hedeman, E. R.: 1969b, *Solar Phys.* **9**, 278.
Gleissberg, W.: 1945, *Observatory* **66**, 123.
Gleissberg, W.: 1952, *Die Häufigkeit der Sonnenflecken*, Akademie-Verlag, Berlin.
Gnevyshev, M.: 1963, *Soviet Astron. – AJ* **7**, 311.
Kopecký, M.: 1960, *Bull. Astron. Inst. Czech.* **11**, 35.
Maunder, E. W.: 1890, *Monthly Notices Roy. Astron. Soc.* **50**, 251.
Maunder, E. W.: 1904, *Monthly Notices Roy. Astron. Soc.* **64**, 747.
Newton, H. W. and Milsom, A. S.: 1955, *Monthly Notices Roy. Astron. Soc.* **115**, 398.
Schove, D. Justin: 1955, *J. Geophys. Res.* **60**, 127.
Švestka, Z. and Simon, P.: 1969, *Solar Phys.* **10**, 3.
Vitinskii, Yu. I.: 1960a, *Soln. Dann.*, No. 8.
Vitinskii, Yu. I.: 1960b, *Soln. Dann.*, No. 9.
Vitinskii, Yu. I.: 1962, *Solar Activity Forecasting*, published by the Academy of Sciences of the U.S.S.R., Leningrad.

Waldmeier, M.: 1935, *Astron. Mitt. Zurich* **14**, No. 133.
Waldmeier, M.: 1957, *Z. Astrophys.* **43**, 149.
Waldmeier, M.: 1966, in G. Righini (ed.), 'Proceedings of the Meeting on Sunspots', G. Barbèra, Firenze, p. 41.
Wolf, R.: 1861, *Astron. Mitt. Zurich*, No. 12.
Zhukov, L. V. and Muzalevskii, Yu. S.: 1969, *Astron. Zh.* **46**, 600 (Translated: *Soviet Astron. – AJ* 1969, **13**, 473).

## Data Sources

*Monthly Bulletin of Solar Phenomena*, Rome Observatory, 1962–1967.
*Quarterly Bulletin on Solar Activity*, International Astronomical Union.
*Royal Greenwich Observatory Bulletins*, Photoheliographic Results, 1955–1961.
*Sunspot and Geomagnetic-Storm Data*, derived from Greenwich Observations, 1874–1954.

# METHODS FOR THE FORECASTING OF SOLAR FLARES

A. B. SEVERNY and N. V. STESHENKO
*Crimean Astrophysical Observatory, U.S.S.R.*

Flares appear only in regions on the sun with strong magnetic fields, mostly in sunspot groups. Two circumstances are of great importance for the forecasting of flares:

(a) the character of the regions producing flares of a certain importance,

(b) the characteristic features of the evolution of active regions.

The most important in practice is the forecasting of proton and other important flares, because of the effects they produce in interplanetary space and in the earth's environment. It is well known that they can represent a real radiation hazard for manned space-flight. Table I gives approximate doses of radiation for several flares in a spacecraft with a protection (shielding) of 1 g/cm$^2$ (Bobkov *et al.*, 1964).

TABLE I

| Flare | Integrated intensity |
|---|---|
| | 10 MeV |
| 25 March 1958 | 1020 rad |
| 10 May 1959 | 700 rad |
| 17 July 1959 | 1080 rad |
| 12 November 1960 | 790 rad |

The lethal dose is 300–500 rad.

Most of the investigations in connection with flare forecast refer to the structure of active regions at different levels of the solar atmosphere, the structure of magnetic fields and their association with flares, and transformations taking place in active regions during a short time-interval (1–2 days) before and after a flare. These investigations showed many important properties of flare-producing active regions. At present we have some quantitative criteria for the possible appearance of flares of certain importance in an active region. It should be emphasized that only those 'parameters' of an active region are considered which characterize this region at the very moment of forecast compiling. The character of the further development of an active region (its growth, complication or simplification) is to be determined (by extrapolation and by considering the similarity with other active regions) on the basis of all data about the evolution of the region since its appearance, up to the moment of forecasting. As a rule, unipolar and well-defined bipolar spot groups offer no difficulty when considering their further evolution. But eastern regions developing on the back

*Dyer (ed.), Solar Terrestrial Physics / 1970: Part I,* 173–181. 

side of the sun present considerable difficulties. Besides that, the formation of new active regions is also extremely difficult to foresee.

Let us consider the results of some investigations that can serve as a basis for quantitative criteria of flare forecasting.

Since 1957, when regular observations of magnetic fields with the Crimean magnetograph started, we have been comparing magnetic fields with flares and examining changes in the configurations and the strength of the magnetic fields associated with flares. It has been found (Severny, 1958) that at the appearance of a flare bright knots occur very close to the neutral line of the longitudinal magnetic field $H_{\parallel}=0$, or even coincide with it; i.e. a flare appears between two polarities in the neighborhood of some (at least three) magnetic hills or spots of different polarities. Next (Severny, 1960) twenty flares were found to appear in regions of magnetic fields with mixed polarities, and high gradients, as

$$\nabla H_{\parallel} > 0.1 \text{ G/km}.$$

A comparison of magnetic field maps, obtained *before* and *after* flares, has shown that flares may lead to a considerable simplification of the magnetic field structure and to a decrease of the field gradients after a flare, down to 0.01–0.02 G/km.

The above-mentioned results, viz. the appearance of flares in regions with high magnetic field gradients and the change of the field structure associated with a flare, provide evidence that magnetic fields of an active region are responsible for flare appearance, but the flare importance and its character are mainly determined by the magnetic field structure and its strength.

Simple energy considerations have shown that there is no other reasonable energy-source for flares than magnetic energy.

The tendency of flares to occur in knots close to the neutral line $H_{\parallel}=0$ has been confirmed. Moreton and Severny (1966, 1968) examined the position of 60 bright knots appearing as first brightenings of 25 flares in the very active region of September 17–24, 1963, and it was found that in 22 flares out of 25 (88%) one of the bright knots was at a distance less than 9″ from the neutral line $H_{\parallel}=0$. Martres *et al.* (1968) found that 67% of 154 bright knots for 32 flares coincided with $H_{\parallel}=0$ within limits of $\pm 10''$. An extensive comparison of flare positions relative to spots in groups with magnetic field determinations before and after flares, was made in the Crimean observatory (Gopasyuk *et al.*, 1963). All important flares studied in this work were divided into 4 groups:

I – proton flares of importance 3 or $3^+$ producing cosmic ray effects recorded with balloons and at ground-level.

II – proton flares of importance 3 or $3^+$ showing no effect on balloon-borne instruments but showing P.C.A.-effect.

III – non-proton flares of importance 3.

IV – non-proton flares of importance 2 or $2^+$.

On the basis of similarity considerations and the data on field strengths, area and position of spots, a laboratory model of the magnetic field of an active region was constructed. The magnetic field gradients near the neutral points of the configuration obtained were measured. Gradients exceeding 0.1 G/km were to correspond to proton flares and all other important flares. Table II shows that the lower the magnetic field gradients are the less important the flares that occur.

TABLE II

| Flare class | Flare type | Number of flares | Field gradient G/km | | Rate of field change (G/days) | Flare duration (days) |
|---|---|---|---|---|---|---|
| | | | Before a flare | After a flare | | |
| 1 | Proton (cosmic rays) | 13 | 0.78 | 0.27 | 400 | >0.17 |
| 2 | Proton (PCA) | 11 | 0.45 | 0.27 | 284 | >0.089 |
| 3 | Importance 3 without PCA | 12 | 0.18 | 0.13 | 83 | >0.071 |
| 4 | 2+ and 2 usual | 15 | 0.054 | 0.038 | 40 | 0.048 |

The quantitative correlation found was used for an experimental flare forecast during some months (about 800 flares of different importance were observed during this period). The flare importance was predicted according to the scale:

(1) No flares or flares of low importance are expected.

(2) Flares of mean importance are expected.

(3) Flares of great importance are expected.

A comparison of the forecast of flare importance with the flares appeared has shown 80% of such forecasts to be correct (see Figure 1).

Other investigators have also come to the conclusion that large gradients of the magnetic field are associated with regions of flare appearance. Avignon *et al.* (1964, 1966) and Caroubalos (1964) have found a statistical correlation between important flares and the ratio $d/D$, where $d$ is the distance between the spots and $D$ is the average diameter of the spots.

This ratio is a kind of characteristic of the mean value of the magnetic field strength gradients in the group.

The forecasters of the Space Disturbance Forecasting Center in Boulder use the H$\alpha$ fine structure as a tool for forecasting solar flares (Knecht, private letter). Since they do not have real-time magnetic field data, they determine the magnetic structure of the active region from the positions of dark filaments, fine fibril structures, plage corridors, bright plages and sunspots. The connection of all these features with magnetic fields was studied by many astronomers (Stepanov, 1958; Martres *et al.*, 1966; McIntosh, 1969; McIntosh and Sawyer, 1969; Smith, 1968).

The forecasters in Boulder have found that in most cases the polarity of a region can be correctly deduced if it contains a large, symmetrical leading sunspot. This assumption is especially valid if the sunspot is located in the western end of the region. The polarities of adjacent areas are deduced by alternating polarities at the border of each longitudinal 'neutral' line. The position of the neutral line is deduced from the location of dark filaments, which, for a part of their length follow the transverse field line, and other elements of the Hα line structure. In those regions where the place of change of polarity is interrupted by complex, bright plage structures, which apparently cross the line of change, active areas apparently occur. When this situation exists the forecasting center assumes that the region is at least mildly complex, and therefore, potentially may yield important flares. However this method is largely a qualitative one.

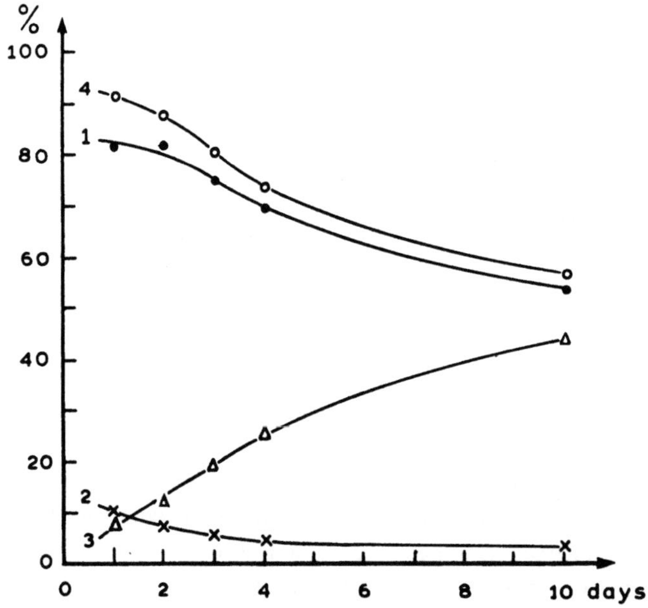


Fig. 1. Justification of forecastings of flare importance. 1: percentage of correct forecastings; 2: overestimated forecastings; 3: underestimated; 4: the sum of correct and overestimated forecastings.

There are many other qualitative confirmations of the fact that the most important flares tend to appear in areas of a very steep gradient of magnetic field. E.g. Warwick (1966) has come to the conclusion that all proton flares occur in spot groups with a δ-configuration according to the Potsdam classification (proposed by Künzel (1960)) i.e. in groups where two or more umbrae of different polarities occur in the same penumbra. This circumstance explains the fact that the most important flares, as was first observed by Dodson and Hedeman (1949), cover the umbra of the nearest spot when developing.

Michard *et al.* (1961) were the first to show that, while flares are closely associated with the neutral lines $H_{\parallel}=0$, they mostly occur in those parts of that line where a strong transverse field ($H_{\perp}\neq 0$) exists.

Regular recordings of the transverse field in active regions made with the Crimean magnetograph showed highly inhomogeneous structures of the transverse field $H_{\perp}$ and a much more inhomogeneous structure of $H_{\perp}$ than $H_{\parallel}$. An examination of the position of flares relative to the $H_\perp$ fields showed (Severny, 1964) the tendency of flares to appear in regions of peculiar behaviour of the transverse field such as a strong rotation of the $H_\perp$-vector, or the 'inclusion' of transverse fields in regions occupied only by $H_\parallel$, or in unipolar regions. Flares also appear in regions where the conjunction of opposite $H_\perp$ vectors takes place, or in 'bifurcated' regions where the directions of $H_\perp$ apparently cross. Hence, in addition to the large gradients of the magnetic field component $H_\parallel$, another important feature – great changes in the direction of $H_\perp$ – must be considered.

Sometimes, as e.g. in the flare of September 20, 1963, the size of the contact region of the magnetic field component $H_\perp$ with opposite directions is only $\approx 5''$ and most flares appear in this region. The probability of flare appearance in these special bifurcated $H_\perp$ regions is $\approx 83\%$, about equal to the probability of flare association with the neutral line $H_\parallel = 0$.

The strong variations of the net magnetic fluxes observed in connection with large flares offer also a possibility to estimate and also to predict the electromotive force due to the change of the net flux with time, or to estimate the possible energies of protons that could be generated by flares.

From the above we may conclude that quantitative methods to forecast flares within the next 1 to 3 days can be elaborated on the basis of maps of the longitudinal and transverse magnetic field components.

It is quite natural that in actual practice one must not only consider the above-mentioned quantitative characteristics of magnetic fields, but also a number of qualitative properties of active regions, such as the character of the magnetic field changes in the region during its development, the presence of rapid and violent motions, the appearance of dark eruptive filaments, the 'turbulent' patchy appearance of the Hα-line, the frequency of the appearance of non-stationary continuous and line-emission ('moustaches'), the brightness of Hα and K-plages, the character of the radioemission etc.

If e.g. the measured gradients of the magnetic field correspond to flares of importance between 2 and 3, and the group continues to develop and may be regarded as an active or eruptive one, the possibility of a flare of importance 3 will be announced. When the group is close to the limb so that gradients of the magnetic field can not be measured, only the complex of the above-mentioned characteristics should be used for the forecast.

Special attention should be paid in this respect to the qualitative data on the character of the evolution of sunspot groups as found in Meudon (Simon, 1968). The birth of a new group close to an old one may give rise to the development of a complex structure group and to the appearance of an important flare, especially if this new group is formed to the west (ahead) of the leading spot. Křivský and Obridko (1969) have found that other spot groups at some distance may exert some influence on the development of the group.

It should be noted that daily heliograms of groups obtained on a large scale and with high resolution, would be of great use for determining the spot group structure, the character of its evolution, and its activity level. For the very complex spot group that caused the proton flare of July 7, 1966, it was shown (Steshenko, 1969) that the filaments in the penumbra of the group repeat the transverse field direction on the whole. Similar results have been obtained by Kuklin and Stepanov (1963) and by Tsap (1964) for simpler spots.

If it would turn out that the direction of the filaments repeats the structure of the transverse field in those special points where flares occur, it would be easier to obtain the observing material necessary to analyze the level of group activity.

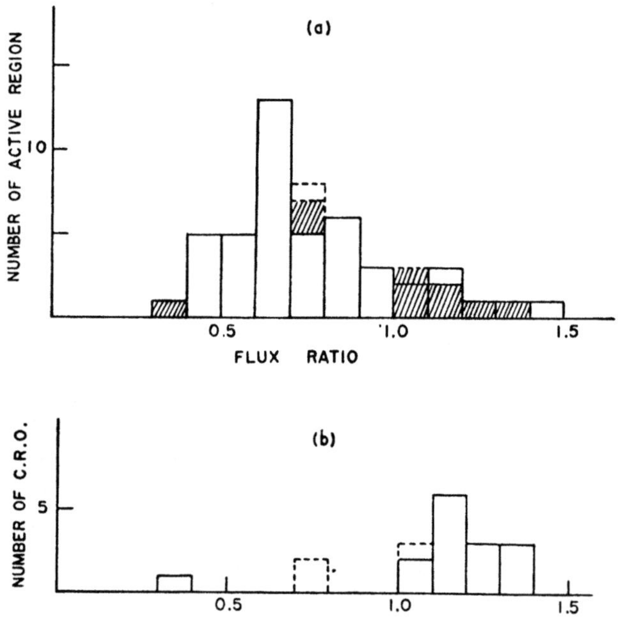


Fig. 2. Illustration of the results of Tanaka and Kakinuma.

It should also be noted that for most great proton and ground-level cosmic-ray flares the inversion line of a spot group was very close to the east-west direction.

Observations of the radio emission of active regions may be important for flare forecasting. So, Kundu (1959) found a high probability of the appearance of flares and bursts at centimeter waves in an active region with small dimensions and a high brightness temperature.

Pick (1961) showed that the probability of a Type IV-burst is very high when the intensity of the *S*-component in the active region exceeds $60 \times 10^{-22}$ Wm$^{-2}$ Hz$^{-2}$. Tanaka and Kakinuma (1964) showed that most proton flares appear during periods when in the active regions the ratio of the flux density of the *S*-component of radiobursts at 3.2 cm to that at 7.5 cm is $>1$. Moreover, most proton flares appear in active regions with a flux density of the *S*-component exceeding 30 units. In Figure 2, after Tanaka and Kakinuma, the distribution of active regions and cosmic ray flares is shown to depend on the ratio of fluxes $F$ (3.2 cm)/$F$ (7.5 cm). In the histogram the parts corresponding to the active regions with cosmic ray flares are shaded.

We should also keep in mind that according to the statistics made at the Crimean observatory about 30% of type-IV bursts and important flares accompanying them are not connected with proton events on the earth (neither balloon cosmic ray increase nor PCA events). Probably this is connected with special propagation conditions in interplanetary space and therefore systematic space observations are so important.

There also appears to be no spectroscopic difference in the visible spectrum between 'proton' and 'non-proton flares' (Koval and Steshenko, 1963). Hence, also taking into account the experimental data on cosmic ray effects from minor flares (McCracken, 1968), we would probably be not far from the truth when assuming that there is no qualitative difference between proton and non-proton flares.

On the basis of research by Kundu (1959) and Pick (1961), the Meudon observatory began in 1962–63 with the forecasting of flares, or rather of the level of spot group activity. Later on they used the criteria obtained by Tanaka and Kakinuma (1964) and the above-cited correlation between important flares and the ratio $d/D$ – which is analogous to the magnetic field gradients. A magnetic classification and qualitative data on the evolution of groups are used as important forecasting parameters at Meudon.

The long or average term forecasting of solar flares is still more difficult than short-term forecasting. Here we can only follow the statistics in the distribution of active regions over the Carrington longitudes on the Sun. Quantitative data for these distributions have been found by Warwick (1965) and Levitsky (1967). The longitudinal distribution of proton flares obtained by Levitsky is given in Figure 3. Flares with powerful Type IV bursts giving long PCA's are indicated by thick shading. Weak shading indicates flares accompanied as a rule by Type IV bursts with an increase of $f_{min}$ but without long complete PCA's. The flares, after which a weak increase of $f_{min}$ was observed, are indicated by dots. Almost half of these flares were not accompanied by Type IV bursts. The histogram clearly shows the presence of a wide longitudinal region, from 140° to 280°, where very important proton flares occur less frequently than in other regions on the Sun. The distinction between these two regions proved to be stable during the whole considered period – about 7 yr. Assuming that the distribution of active regions during this cycle of solar activity is similar to that of the last cycle, one can choose the time periods for which the probability of flare appearance is relatively low or high. However there is no way of forecasting the time period for the appearance of proton spot groups for an active longitude. Švestka (1968) notes that taking into account the frequency of formation of complex spot groups capable of producing proton flares, it is impossible, on the basis of statistic data, to predict days free of proton danger during three years close to the solar activity maximum (i.e. during 1968–70 for this cycle).

To forecast the proton danger of flares it is necessary to take into account the conditions of proton propagation in interplanetary space as noted before.

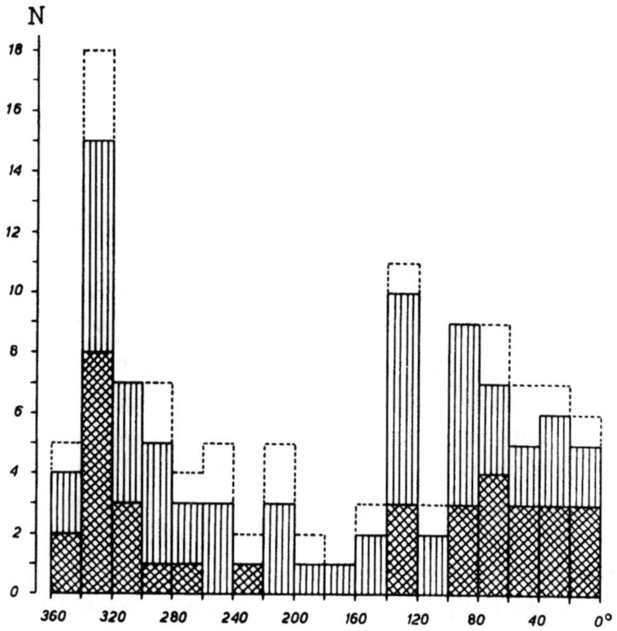


Fig. 3. The longitudinal distribution of proton flares.

On the basis of statistical research Levitsky and Vladimirsky (1969) have found the so-called zone of avoidance for solar protons between heliographic latitudes of $\pm 2°$. In this region the probability of appearance of solar protons is appreciably reduced. This circumstance can also be used for forecasting the proton danger of flares due to propagation conditions as well as the well-known east-west asymmetry on the solar disk for proton events.

## References


Avignon, V., Martres, M. J., and Pick, M.: 1964, *Ann. Astrophys.* **27**, 23.
Avignon, V., Martres, M. J., and Pick, M.: 1966, *Ann. Astrophys.* **29**, 33.
Bobkov, V. G. *et al.*: 1964, *Radiacionnaja bezopasnost pri kosmiceskih polëtah*, Atomizdat, Moskwa.
Dodson, H. W. and Hedeman, R.: 1949, *Astrophys. J.* **110**, 242.
Caroubalos, C.: 1964, *Ann. Astrophys.* **27**, 333.
Gopasyuk, S., Ogir, M., Severny, A., and Shaposhnikova, E.: 1963, *Izv. Krymsk. Astrofiz. Observ.* **29**, 15.
Koval, A. N. and Steshenko, N. V.: 1963, *Izv. Krymsk. Astrofiz. Observ.* **30**, 200.
Krivsky, L. and Obridko, V.: 1969, *Solar Phys.* **6**, 418.
Kuklin, G. and Stepanov, V.: 1963, *Soln. Dann. Bull.* **1**, 55.
Kundu, M. R.: 1959, *Ann. Astrophys.* **22**, 1.
Künzel, H.: 1960, *Astron. Nachr.* **285**, 271.
Levitsky, L.: 1967, *Izv. Krymsk. Astrofiz. Observ.* **37**, 137.
Levitsky, L. and Vladimirsky, B.: 1969, *Izv. Krymsk. Astrofiz. Observ.* **40**, 161.
McCracken K. G.: 1968, XIth COSPAR Meeting, Tokyo.
McIntosh, P. S.: 1969, *Ann. IQSY* **3**, 40.
McIntosh, P. S. and Sawyer, C.: 1969, *Ann. IQSY* **3**, 169.
Martres, M. J., Michard, R., and Soru-Iscovici, I.: 1966, *Ann. Astrophys.* **29**, 249.

Martres, M. J., Michard, R., Soru-Iscovici, I., and Tsap, T.: 1968, in K. O. Kiepenheuer (ed.), 'Structure and Development of Solar Active Regions', *IAU Symp.* **35**, 318.
Michard, R., Mouradian, Z., and Semel, M.: 1961, *Ann. Astrophys.* **24**, 54.
Moreton, G. and Severny, A. B.: 1966, *Astron. J.* **71**, 172.
Moreton, G. and Severny, A. B.: 1968, *Solar Phys.* **3**, 282.
Pick, M.: 1961, *Ann. Astrophys.* **24**, 183.
Severny, A. B.: 1958, *Izv. Krymsk. Astrofiz. Observ.* **20**, 22.
Severny, A. B.: 1960, *Izv. Krymsk. Astrofiz. Observ.* **22**, 12.
Severny, A. B.: 1964, *Izv. Krymsk. Astrofiz. Observ.* **31**, 159.
Simon, P.: 1968, Lecture at the Crimean Astrophys. Observ.
Smith, S. F.: 1968, in K. O. Kiepenheuer (ed.), 'Structure and Development of Solar Active Regions', *IAU Symp.* **35**, 267.
Stepanov, V. E.: 1958, *Izv. Krymsk. Astrofiz. Observ* **20**, 52.
Steshenko, N. V.: 1969, *Izv. Krymsk. Astrofiz. Observ.* **39**, 245.
Švestka, Z.: 1968, *Solar Phys.* **4**, 18.
Tanaka, H.: 1969, *Proc. Res. Inst. Atmosph. Nagoya Univ.* **16**, 203.
Tanaka, H. and Kakinuma, T.: 1964, *Rep. Ionosph. Space Res. Japan* **18**, 32.
Tsap, T. T.: 1964, *Izv. Krymsk. Astrofiz. Observ.* **31**, 200.
Warwick, C.: 1965, *Astrophys. J.* **141**, N2.
Warwick, C.: 1966, *Astrophys. J.* **145**, 215.

GPSR Compliance

*The European Union's (EU) General Product Safety Regulation (GPSR) is a set of rules that requires consumer products to be safe and our obligations to ensure this.*

*If you have any concerns about our products, you can contact us on ProductSafety@springernature.com*

In case Publisher is established outside the EU, the EU authorized representative is:

Springer Nature Customer Service Center GmbH
Europaplatz 3
69115 Heidelberg, Germany

**Batch number: 10620881**

Printed by Printforce, the Netherlands